HEART OF A LION

THE COURAGEOUS JOURNEY OF JANET LEE SIMPSON

BY

CONNIE DONALDSON

Copyright ©2006 by Connie Donaldson

ISBN 0-9759818-5-7

All rights reserved.

Cover Design by Connie Donaldson

Published and distributed by:
High-Pitched Hum Publishing
321 15th Street North
Jacksonville Beach, FL 32250

Contact High-Pitched Hum Publishing at www.highpitchedhum.net

No part of this book may be reproduced or transmitted in any form or means, electronic or mechanical, including photocopying, recording, or by any information storage and retrieval system, without permission in writing from the publisher.

ACKNOWLEDGEMENTS

The author wishes to thank Janet Lee Simpson for all the shared hours when she opened up her home, her past, and her heart so honestly and hospitably. Enormous thanks are due to Kathryn James for her always-insightful help with facts and stories from her treasure trove of memories, and to Rob Simpson. Special appreciation to Clara Schultz, Carol Pate, Barcher Threshie and Beverly Joy Combs for support both practical and emotional, and to the Sierra bunch (you know who you are). Thank you Pamela DeNeuve for helping to inspire this project. Lastly I thank my husband and best friend Wharton Donaldson for making it all possible, believing in me, and acting as unpaid proofreader and literary critic, a role he is irritatingly good at.

DEDICATIONS

To the memory of my parents O'Dear and Harry and my sister Gail, who taught me so much yet fell victim to the disease of alcoholism. I wish you were here to share the sunlight of recovery.
- Connie Donaldson

For my sister Kathryn, who was there through it all, and my niece Pam who with her courage shares the lion's heart. And for my son Robbie, the wind beneath my wings.
- Janet Simpson

AUTHOR'S NOTE

Historical data, facts, names, and other details have been reported with as much veracity as possible based on my sources. Any errors are my own. Where privacy might be violated I have changed the names of those concerned.

HEART OF A LION
THE COURAGEOUS JOURNEY
OF JANET LEE SIMPSON

BY
CONNIE DONALDSON

INTRODUCTION

From the first time I heard Janet Simpson's story, I knew it needed to be told, and that I wanted to be the one to tell it.

On the surface, Janet's life and my own have almost nothing in common. She is a woman of action, a sportswoman, comfortably at home behind the controls of an airplane, walking five miles on the beach on Turtle Patrol, competing in a golf tournament. None of those things are my forte (to put it mildly). Yet, Janet and I share a deep sisterhood of mutual understanding that transcends age, interests, and history. We both know what it is to struggle with a disease which does not submit to all the willpower and good intentions in the world. We both know the shame and secrecy that comes with the territory. And we both know the inexpressible joys and rewards that come with being in recovery.

Our friendship deepened as we spent afternoons at her house, sitting on the open lanai watching the birds throng to her feeder or catching a glimpse of Beau the neighborhood gator in the pond. Together we poured over photo albums and scrapbooks, stacks of newspaper clippings, and mementos from her incredible past. As Janet spoke the years rolled back and I was there with the dimpled tomboy tussling with her pet lion cub on the floor. I was once a tomboy too, and with Janet (whose spirit is truly ageless) I could feel that girl still within me. She had found a great pal.

Of course, I would usually be accused of "making her cry" before the session was over, as Janet reconnected with old, tender feelings. But it was all to the good.

It was healing for me to be with a recovering woman of my own mother's generation who would talk openly and without shame about the disease of alcoholism. I lost my father, mother, and younger sister to its ravages. I was spared, and in gratitude I try to share hope with others. So does Janet.

Noble motives aside, I love a good story and know one when I hear one. Janet's story combines high drama, history, extravagant characters, conflict, exotic settings, and deep emotion.

So here's to following the instinct and the heart, as I did that night when I decided I wanted to help Janet's dream of having her story told become a reality. It has been an honor and a privilege, as well as lots of fun.

1

IN THE HEART OF THE LION

In the lion's heart dwells daring, danger, the possibility of destruction. The remarkable life of Janet Lee Hutchinson Simpson took her through the heart of darkness to the place where the lion, in harmony and peace, lies down with the lamb at last. This is her story.

The tiny blond woman with the shining blue eyes glanced nervously around the meeting room. A sea of encouraging faces surrounded her, yet Janet's heart was beating fast. This woman of action struggled to find the words to tell the story of her journey, of hitting bottom from alcoholism at age 75 and her struggle to build a new, sober life to a roomful of people. At least they, too, were recovering alcoholics.

Words that had come easily to the young Janet when she appeared on Broadway or opposite such stars as Gary Cooper on the radio now stuck in her throat. How could she ever describe the twisting path that had led her here?

Janet never ran from a challenge. She took a deep breath and slowly the words started to come.

"I was brought up to set goals for myself and I was able to meet every single goal I ever set," she began. "I wanted to learn to fly, and I did. I wanted to be a radio star, and I became one. I wanted to train racehorses and play championship golf and tennis. I did all those things."

In the next hour, Janet's listeners sat transfixed by her halting words as scenes from an incredibly rich and adventurous life unfurled.

The author was one of those rapt listeners. Janet's was a story that needed to be told, and in that little room the seed for this book took root that very night.

* * *

September 11, 1932.

At age seven Janet Lee Hutchinson, youngest member of America's Flying Family, found herself marooned on a tiny island in the Arctic Ocean with very little hope of rescue. The Hutchinson expedition of father George, mother Blanche, sisters Kathryn and Janet Lee and five crew members had crashed off the coast of Greenland in an attempt to become the first family to cross the Atlantic by air.

Life magazine's Cover Photo of the Year showed Janet Lee, Kathryn and Blanche in silhouette standing on the edge of a boulder watching the tattered hull of their Sikorsky Amphibian sink into iceberg-flecked waters.

Janet didn't remember feeling fear, but she did remember what came next. George Hutchinson wept as he watched his dreams swallowed up by the dark, storm-tossed waters. Tears flowed down his icy cheeks as the plane slowly disappeared. Finally only one upended wingtip remained visible. It was the only time Janet had seen her indomitable father weep, and that moment stayed with her all her life. In that single display of vulnerability the bigger-than-life-figure of her father became life-size.

As the Sikorsky disappeared, Janet and the little group found themselves truly stranded, cut off from any communication with the outside world.

The near-fatal Greenland crash sent America rushing to their radios to learn of the fate of the two little girls, their beautiful stylish mother, and the dashing young father with his leather flying togs and white silk scarf. Newspapers and magazines around the world covered the crash, and commentators such as Lowell Thomas debated the question of parental irresponsibility for taking small children on such a dangerous trip.

George Hutchinson answered these critics with characteristic decisiveness:

We would never leave anyone behind. None of us would ever consider taking such a journey separately. We are a unit; where one goes, all go.

This belief was the core religion of George Hutchinson, a charismatic and determined man who created, out of his own love affair with the fledgling phenomena of aviation and by sheer force of will, America's Flying Family. Starting in 1926, when Kathryn was only 3 years old and Janet Lee 18 months, George and wife Blanche packed the little girls into the cockpit of his single-engine airplane with as little fuss as when other families piled into their Model T for a Sunday afternoon drive. From this simple beginning, George's expanding dreams catapulted the Hutchinsons to national and even international notoriety during the dark days of the Depression. Like Charles Lindberg and Amelia Earhart, they became early media darlings of aviation.

No, George didn't dream small. This was the man, the adventure, and the time that shaped Janet Lee Hutchinson.

The Flying Hutchinsons were to meet with two American presidents and with every governor of the then-48 states on a goodwill tour to promote aviation (with a mascot lion cub in tow), and an All-Nations Flight to promote world peace. After the Greenland trip the family would perform live on their own radio program dramatizing their adventures, and appear onstage nightly in their own vaudevillian show at Loew's Theater on Broadway.

Janet Lee, the "Shirley Temple of the Airwaves," would go on to perform live on NBC radio with such stars as Freddie Bartholomew (her first love) and Gary Cooper.

In World War II Janet took to the skies in the pilot's seat as part of the Women's Airforce Service Pilots or WASP, the first women to fly military planes in the service of their country. Janet's love of horses led to a career as a racehorse owner and trainer.

* * *

Janet's words wove images of a life her listeners had difficulty even imagining, a life lived large and daringly. Here, obviously, sat a woman of great willpower and determination.

Yet, as Janet's voice grew softer still and she spoke of the dark days of escalating drinking, of a time when she lay near death in a hospital bed as the result of a car crash, her listeners grew even more intent, leaning forward in their seats.

They were beginning to identify with the *feelings* Janet shared, for they remain essentially the same for every alcoholic. The emotional climate in the room shifted as each person there recalled the shame and secrecy that had once governed their own lives and brought them to these rooms.

Janet paused. Tears had sprung suddenly to the blue eyes.

"Like I said, I met every single goal I ever set. I always claimed I never knew the meaning of fear, but the one thing I couldn't do was save myself from becoming an alcoholic. That made me very afraid."

2

ROOTS & VINES

Janet's parents George and Blanche Hutchinson were highly unusual people in their own right. Self-contained Blanche served as a perfect foil to the dynamic George. Each brought colorful family histories and strong personalities to the mixture that was to become America's Flying Family.

George Hutchinson came from British stock. His grandfather had been a "rectifier" or taster in a whiskey distillery. His father was a British sea captain, who sailed to Maryland and was so impressed with America he wired his employer he would not be returning but would make the United States his home. He trained his First Mate to take over for the return voyage. From that point, everyone in the Hutchinson clan was born in Maryland.

Grandmother Hutchinson, George's mother, came from Manchester. An actress, she was part of a traveling Shakespearean company that traversed England. Her future husband sent for her from America and, alone, she made the trip across the Atlantic to marry him. Grandfather Hutchinson was gifted. He composed music and painted. His son George would demonstrate the same kind of easy creativity in almost anything he turned his attention to.

The Hutchinsons settled in a row house in a middle class neighborhood outside Baltimore. They were a family of high moral standards, and staunch teetotalers.

George had two siblings. His younger brother Leonard was a natural follower, quiet and sweet in nature, the opposite of his assertive older brother. His beautiful sister May was killed at 16 in a fall from a trolley.

Janet always felt somewhat neutral about her father's family, who lived nearby. After her husband's death, Grandmother Hutchinson lived with Janet's family six months out of the year. Her bland personality did little to bring her little granddaughters close to her. She was extremely overweight, and on one memorable occasion had to be pried out of the bathtub by the local fire department.

* * *

Blanche's family, the Delchers, boasted a larger-than-life Auntie Mame figure in her mother BB. This stood for Big Blanche, while of course Janet's mother was known as Little Blanche. Kathryn's first name was also Blanche. When it came time to name her youngest daughter, Little Blanche rebelled. Determined to buck tradition, she christened the baby Janet after Janet Gaynor, her favorite movie star.

BB's extravagant household was the diametric opposite of that of the quiet Hutchinson clan. Like Mame, BB dangled an ivory cigarette holder from her long, bejeweled fingers and feigned a British accent. She never entered a room without pausing dramatically in the doorway to make sure all eyes were upon her. BB reigned as matriarch supreme of the Delcher clan.

BB had inherited a fortune following her own parents' simultaneous death in a freak automobile accident when she was very young. This left other family members in a bitter fight for her custody.

Blanche lived large, as one to the manor born. She owned a fleet of cars and a summer estate in Riviera Beach, Maryland, where the family visited summers and holidays.

Her first marriage, to the husband of her four children, did not last. She divorced him (quite unheard of at that time and within that social class) and remarried twice. Her first husband, however, remained a good friend and very much still on the scene. Granddaddy Delcher was much beloved by Janet and Kathryn.

BB's second husband, Judge Bond, came from the family for whom Bond Street and the Bond Building in Baltimore were named. After his death she married Norman Stadiger, who eventually became known as "The Judge" as well, even though he was really a magistrate in Ann Arundel county where he and his wife summered in a luxurious waterfront estate.

The elegant product of the best schools, BB made three trips around the world in her lifetime, the first just after graduation from finishing school. This traditional Grand Tour was followed by an excursion that occurred "between husbands". Wooed in Italy by a besotted Count, she brought him back to the States. The family, however, turned thumbs down and the nobleman soon found himself on a boat headed back to his native land.

A true Southern Belle, BB always presented a fashionable and bejeweled exterior to the world. She loved the spotlight. At parties she played the piano, sang, danced and recited poetry at the drop of a hat. Her playing and singing were both exceptional, and her guests looked forward to the moment when BB would break into a spirited version of *Dixie*.

Strangely, considering her own unconventional personal life, BB became the fierce protector of the family history and reputation. She became especially incensed should anyone make a reference to the "Civil War".

"It was the War Between the States, and don't you forget it!" she'd insist with a steely-eyed glare.

She had reason. Her family had found themselves in a peculiar position in the Civil War years. The state of Maryland, situated on the cusp between North and South, was important to both sides. Early in the conflict, Yankee troops occupied Baltimore. Southern Marylanders sided with the Rebels. BB's family was spread across the state, half ending up fighting for the Confederates and half for the Yankees.

BB outlived all her husbands. Her four children were Blanche, Kathryn, Namuth, and Harry. Both Namuth and Harry were casualties of the first World War. Namuth was buried in France, and Harry died young after being gassed.

* * *

Quiet, poised Little Blanche grew up in a household that revolved around her charismatic mother. It was good training for the years to follow, when she held to her calm center in the whirlwind that surrounded life with her dynamic husband.

In BB's household, the matriarchal word was law. She was used to having her own way. But all that was to change, when her young daughter Blanche met the man who was to sweep her off her feet and into a remarkable new life that outdid anything her mother could come up with in terms of theatricality and drama. His name was George Hutchinson.

* * *

George, at that time a student at Johns Hopkins, was an appealing young man with a full head of thick, wavy auburn hair and charisma to burn. A favorite with the ladies, young George was outgoing, charming, and full of fun. His spectacular gift of gab served him well in fundraising and public speaking in later life.

Like BB George insisted on being the center of attention. Whenever anyone around him told a joke, George never laughed. But the next day he would repeat it himself as if it were his own and crack himself up along with his audience.

Gifted with a beautiful singing voice and a great sense of rhythm, George regularly entered the dance contests that were all the rage in the Twenties. He entered one with a beautiful young flapper who aspired to become a professional dancer, and they won hands down. A few years later she graced the silver screen as Joan Crawford.

This was all part of the "unconventionality" that BB deplored in her daughters' suitors, if not in her own person. Her youngest daughter Blanche, seventeen, was still living at home when George appeared on the scene. From the first introduction, BB rejected him as a suitable candidate for her cherished daughter's hand. She did not think much of his prospects, and viewed the Hutchinsons as an undistinguished family not up to her elevated social standards.

There was much more to George Hutchinson than charm and high spirits, however. A young man of uncompromisingly high principles, he never drank, smoked, or swore. His daughters remembered, years later, that in the heyday of The Flying Hutchinsons fame he had resolutely turned down a then huge sum of money offered to endorse cigarettes. He didn't believe in smoking, and held firm.

Unimpressed, BB forbade her daughter to see George. Perhaps part of her disapproval lay in the fact that she was totally unaccustomed to sharing the limelight. Wherever Geroge Hutchinson went, he commandeered center stage. Unlike everyone else around BB, George refused to be intimidated or unduly impressed by her drama. This was perhaps the unkindest cut of all to BB's pampered ego.

Actually the two were much alike in personality.

But Little Blanche was smitten, and from the day she met George she never looked back. Uncharacteristically, she defied her mother and moved out to live with her sister Kathryn who worked at Crown, Cork and Seal (the largest employer in Baltimore at that time.)

Soon after, Blanche and George eloped. BB was furious, and icy distance prevailed until the birth of her first baby. When little Kathryn arrived on the scene, BB melted.

* * *

Perhaps Blanche's exposure to her dominant mother had something to do with her ability in later years to adapt to the most unconventional and daunting of George Hutchinson's plans and enterprises. According to her daughters, Blanche approached perfection. She was sweet natured, kind, giving, and slow to anger. She loved her husband completely, and if she had doubts about some of his riskier ventures, she managed to tame her fears and follow wherever he led. She usually kept her qualms to herself, except in extreme situations, when she could be quietly firm.

The love story of Blanche and George was to be an enduring one. Their daughters never saw them fight, or even argue. George openly adored his petite wife. Often he would look up when his wife entered a room and proclaim:

"Girls, just look at your Mother. Isn't she beautiful?"

Blanche Hutchinson had the oval face, large limpid eyes, and porcelain complexion of a true English beauty. She was careful to protect that skin from the sun, never gardening without donning a big hat and gloves.

Still, the usually sweet-natured Blanche had a temper if "riled" and was fiercely possessive of her handsome husband. If an admiring woman came on too strong, or George showed more interest in another woman than she deemed proper, Blanche didn't hold back.

On a family outing to the beach Blanche noticed a sunbathing woman staring at George a little too intently for her taste.

"Get out of here, you!" she yelled. And the woman got out of there, fast.

George had been working part time as a bank teller and attending school, but quit college in his sophomore year when Kathryn was born. He went to work for Lee Tire and Rubber (the only regular office job the girls remembered their father ever holding). BB, from this point, would offer financial help to the struggling family at various times.

* * *

Janet Lee Hutchinson was born on April 3, 1925 in the Women's Hospital in Baltimore. Janet was irrepressible from the start, an active and courageous little tomboy who grew up to fill the place of George Hutchinson's missing son.

The youngest Hutchinson was towheaded, with shining blue eyes, dimples that wouldn't quit, and a wide and ready smile. Like her father she was an extrovert and loved the limelight. Like him she excelled at singing, dancing, and elocution. From earliest days, Janet's passions were fierce and when she loved she loved hard and loyally. She adored animals, and they responded to her in kind.

Auburn-haired Kathryn was more like her mother. Quieter, introspective and artistically inclined, Kathryn was always devoted to her studies. She acutely observed all that went on around her, but unlike her little sister she shrank somewhat from being in the public eye.

Lee Tire and Rubber transferred George to Abingdon, Pennsylvania, off the Main Line in Philadelphia. The family moved into a roomy house with twin chimneys and a big yard that offered a wonderfully woodsy place for the girls to play.

Janet's earliest memory was of falling down the steps at age five. She loved to ride up and down in the dumbwaiter in the big old dining room. At age six, she broke her leg sledding. Attempting a "triple decker" with Janet on top of Kathryn and her Dad on the bottom, her leg got caught in the runner and she was dragged downhill.

It didn't faze her for long. "I had fun on those crutches!" she remembers. That same year, Kathryn was hit by a truck and broke a few ribs, but wasn't seriously hurt. "Our family never made a big deal out of accidents," Janet recalled.

By all accounts, Kathryn and Janet both had wonderful childhoods. The girls were surrounded by an extended family system in the Baltimore area, and they adored their young, funloving parents.

It was great fun being a child in the Hutchinson clan. You never knew what adventures were just around the next corner. Yet, the bond of this unusually close-knit family held the sisters in at least the illusion of total safety and security.

3

ON THE RUNWAY

George had always been attracted to the drama of the racetrack and the thrill of gambling. One lucky day, out of the blue, a bet paid off and he came home in triumph some $8,000 richer. He determined he would use the money to do what he'd always longed to do. He would learn to fly. And once the flying bug bit George Hutchinson, life would never be ordinary again for the little family.

He bought his first plane and took flying lessons. On training flights, he loved to buzz the family home by flying between the two big chimneys, much to the dismay of the neighbors. Upbraided by an embarrassed Blanche, he'd promise not to do it again, but the temptation always proved too great.

Eventually George bought a big, empty parcel of land outside Philadelphia with another investor, determined to create a private airport. The beginnings were primitive. Nothing was paved. There was one runway, no lights, and one hangar. Grandaddy Hutchinson, now retired, came to Philly and stayed with the family, acting as office manager. It was a start.

Hutchinson had been intrigued with the romance of aviation from its beginnings, when pioneers such as Charles Lindberg and Amelia Earhart began to capture the public imagination. He managed to meet them both, becoming friendly with Earhart's publisher husband, George Putnam. Although Lindberg's fame provided George with a blazing trail to follow, he came to disdain Lindberg's politics in the years before WWII when the famous aviator praised the superiority of the Nazi air force.

Back when people were still afraid to fly, George set out singlehandedly to change the way the average American viewed aviation. Unlike Lindberg, who felt that the big future for the industry lay in air mail delivery, he had a more ambitious vision from the beginning. He wanted to publicize air travel as a safe and reliable mode of transportation for the typical American family.

In addition to charter service, plane sales, student instruction and air taxi work, George began to do air shows and stunt flying on the weekends to attract customers and to publicize the airport. Big crowds would gather to witness aerial acrobatics, parachute jumping, sky writing, night flying and formation flying. Six to ten planes were kept constantly in the air on weekends and holidays.

Ever the entrepreneur, George came up with a scheme to get past the hesitancy, or downright timidity, of the wealthy businessmen he wanted to attract as customers.

He hired a bevy of chorus girls to go up with his clients on short sightseeing flights. Because of the tight quarters, the girls had to sit on the passengers' laps. Business boomed as enthusiastic gentlemen lined up to book flights.

* * *

The name "Flying Family" was wished on the Hutchinsons by the press. Their first flight together as a family, in 1926, set the precedent. Janet Lee was eighteen months old, and Kathryn not quite four.

Hutchinson got past the reluctance of intimidated wives and mothers who were afraid to fly, and shuddered at the thought of their children doing so, by taking little Kathryn and Janet along. With no more fanfare made of it than a ride in the family auto, flying soon became mundane to the girls. Being in the air and feeling safe there became as natural as breathing.

By the time the sisters were six and eight years of age, they had logged more time in the air than the average transport pilot of that time.

The result? "I really didn't learn what fear is supposed to feel like," Janet stated decades later.

Understanding the character of George Hutchinson is the key to understanding the unique lifestyle of the Hutchinsons. He was an unconventional visionary with a creative and inventive mind. His dreams were big, and required a worldwide stage for their execution. Since he remained the consummate family man, his wife and daughters would just have to go along for the ride.

* * *

Onto what national stage was the drama of the Flying Hutchinsons to be launched? As George began to spin out his particular version of the American Dream, all about him other Americans were losing theirs. Panic had seized America on October 24, 1929 - "Black Thursday"- when the stock market crashed. Banks closed. Unemployment soared to 25%, as the median family's income plunged by some 40%, down from an average of $2,300 to $1,500 annually. Almost overnight, a depression engulfed America. Survival became the keynote as Americans questioned the very foundation of the beliefs they lived by.

Herbert Hoover was President. Shantytowns called Hoovervilles sprang up around the country, populated by the down and out and the suddenly homeless. Hoover felt relief should be left to the private sector. All this would shift, of course, on the winds of change that swept in Roosevelt's New Deal. By the end of the Thirties, dozens of new federal programs and commissions would regulate investments, banking and business in this country.

Even in the darkest days of the Depression, people wanted to escape the dreariness and seeming hopelessness of their daily lives. The Thirties were the Golden Age of both radio and movies. It was the era of matinee idols. Clark Gable, Betty Davis, Greta Garbo, Errol Flynn lit up the Silver Screen. Audiences fell in love with little Shirley Temple and laughed at W. C. Fields, the Marx Brothers, and Bob Hope. Family radios brought the wide world into American living rooms. Roosevelt's Fireside Chats calmed fears and talked Americans through the Great Depression and onto the battlefields of World War II.

In what was surely one of the greatest horse races in history, Americans in the thousands sat glued to their radios as the maverick race horse Seabiscuit ran against the heavily favored Triple Crown Winner War Admiral at Pimlico on November 1, 1938. When Seabiscuit won, it was a triumph for the underdog and an affirmation of hope badly needed by a country battered by almost a decade of grim struggle.

Fortunately for the American spirit, there were plenty of both heroes and heroines in that decade between the World Wars. On almost every cultural front, outstanding achievements illumined the darkness. Included in this roster of the greats were Antarctic explorer Richard E. Byrd, scientist Albert Einstein, sports figures Jesse Owens and Babe Diedrickson, the homespun philosopher Will Rogers.

And then there were the aviation greats. Charles Lindbergh made aviation history in his one-man flight across the Atlantic in 1927. From that point on he became the most popular man in the United States. Amelia Earhart became the second person and first woman to fly solo across the Atlantic in 1932. America was rocked in 1932 as well by the kidnapping of the Lindbergh's baby son from their New Jersey home. The nation held its collective breath as a desperate search ensued. Months later, tension turned to horror and grief when the baby was found murdered.

Coincidentally, the first nationally publicized flight of America's Flying Family would take place that same year. Sobered by the Lindbergh case, Blanche and George Hutchinson made sure their little girls were never left alone or unprotected.

Certainly, the exploits of the Flying Hutchinsons would not have captured the attention of the American people so readily were it not for the popularity of radio. Broadcasts lent a dramatic, emotionally involving quality to unfolding events.

Although Lindbergh and Earhart had done much to ignite popular interest in the fledgling aviation industry, it remained the province of the highly daring or the highly wealthy in most people's minds.

In the 1930s, a new generation of low-wing, streamlined, all-metal airplanes were rolling off the production lines, and the United States was in a position of world leadership in aeronautics. Private companies first started delivering air mail. Most American airlines trace their lineage back to these contract mail carriers. Paying passengers were still few and far between.

On a darker note, it was the airplane that gave birth to total war during the years 1939-1945. When the Flying Family took to the skies, aviation was still in its golden age, scarcely aware itself that the air age had truly begun. There would be no turning back.

* * *

George's first big sponsored aviation enterprise came in 1929 when he attempted, without the family, to fly nonstop from Baltimore to Los Angeles in his single-engine plane *The City of Baltimore*. Trouble came just shy of Wichita, Kansas when the arrangements he had made for in-flight refueling (a novel concept in that day) fell apart. He returned home undaunted and with bigger fish to fry. The passion to promote aviation by becoming the first family to take to the skies burned unabatedly.

It is unclear precisely when George Hutchinson conceived the idea of a good will tour of all 48 states to promote aviation. What *was* clear was that he was moving toward this vision when he decided to sell both the airport and the family home and move to New York City. Proceeds from the sale wouldn't begin to cover the kind of expenses his big plans required. That meant finding a sponsor. New York, he knew, was undoubtedly where the big money lay to fund the kind of enterprise he had in mind.

Soon the little family was comfortably ensconced in Manhattan's Governor Clinton Hotel, across the street from Pennsylvania Station. It would be easier for Blanche and the children to live at the hotel while George traveled on his fundseeking endeavors.

* * *

On Christmas Day, 1931, six-year-old Janet's life was about to change in a most exciting way.

It had been a disappointing morning. The little girls missed their father, who had been away on a trip to Wayne, Indiana to secure a new airplane and fly it home. Now a bad snowstorm threatened to keep George from sharing Christmas with his waiting family.

Hopes ignited with a telegram delivered to Blanche at the Governor Clinton:

Merry Christmas to my three sweethearts. Meet me Newark Airport about noon. Bring children, have great surprise. Don't worry.   Love, George.

Blanche and the girls *had* worried, but George's determination had won through once more. He had piloted his new Stinson through heavy snows from Indiana to New York, largely flying blind. In those days pilots relied principally on sight or ground orientation for direction; only a compass and directional gyro aided navigation.

At the airport, the joy of the reunion was exceeded only by the magnitude of the surprise their father had waiting for them.

* * *

When Janet Lee met the little lion cub for the first time, it was love at first sight.

It had been hard for the children to contain themselves as the family gathered around the shining new Stinson. The girls' excitement was not for the new plane, but for what was inside.

"What is it, what is it?" begged Janet Lee, wriggling in her father's arms and trying to see more clearly through the airplane's small windows.

"What's what? I don't see anything," George Hutchinson had always loved to tease. Janet and Kathryn peered up at their father uncertainly.

"The surprise, Daddy, the surprise like you promised!" Kathryn urged.

Winking, George finally pulled open the plane's door. "Now go slowly, girls. Don't scare him," he cautioned.

Peering into the dim cargo hold of the plane, the children saw what looked like a large cat looking back at them from inside a wooden cage.

Janet turned to her father. "But it's just a big cat!"

George chuckled. "It's a cat all right. It's an African lion cub."

When George opened the crate, the cub stood dazed and blinking in the unaccustomed brightness.

Janet's heart beat fast as she tried to take it all in. She saw a comical, furry face and stocky body, tawny as Savannah grass and with paws impossibly too large for its frame. Two yellow eyes with pupils like black diamonds stared right back at her.

Tentatively, the little cub reached out a raspy pink tongue and licked Janet's extended hand.

"He likes me!" she exclaimed. Janet Lee was jumping up and down with excitement. A bit more cautious, Kathryn and Blanche came closer, fascinated as well by this newest addition to the Hutchinson clan.

"It's really a She, honey, but we'll call him He" replied George.

Henceforth, the Flying Family would be joined on their travels by their new mascot "Governor" the lion, the he who was really a she.

* * *

Life at the well-appointed hotel with the big lobby was an adventure for the children from the start. But after the lion cub joined them, things became much more interesting.

"Did you say your prayers, girls?" asked Blanche on Christmas night as she tiptoed in to tuck the little girls in for the night.

"We sure did." Two innocent round faces peered up at her.

A suspicious lump at Janet's feet shifted and grunted softly.

"What's this?" Blanche threw back the covers, knowing of course what she would find. Governor blinked sleepily up at her from the foot of the bed.

"Mother, you woke him up!" Janet protested.

Janet and Governor had bonded instantly, and the little lion followed her everywhere. The bed the Hutchinsons had arranged for her atop soft blankets piled in the porcelain bathtub couldn't match the appeal of Janet's bed. Governor loved to snuggle deep under the covers.

Mornings too took on a new aspect after Governor became a part of the clan.

"Ow! That tickles!" Janet burrowed further under the covers. Governor was licking her head enthusiastically, ready for her sleepy playmate to wake up. Janet rolled out of bed for a brief tussle with Gov before breakfast.

Governor never used her claws as she and Janet wrestled playfully each evening, rolling around on the carpet until both were tired out. The lion never had to be declawed. Governor's gentle and sweet personality, accomodation to walking and heeling on a leash, and adaptability to her surroundings made the pet fairly easy to live with, even in a hotel.

Like all cats, Governor wasn't fond of water. Her first scrubbing in the hotel bathtub elicited howls of protest that sounded like she was dying a slow and torturous death. She liked the drying off part, however, and would cooperate by rubbing her head in the towel. Then came the nail trimming. This wasn't a favorite, either.

The care and feeding of a growing lion was a complicated business. The whole family took turns feeding Governor, to facilitate bonding. Any animal becomes attached to those who feed it regularly, and it was important for the lion to form a firm loyalty to all the Hutchinsons. A carefully regulated diet of cod liver oil, mineral oil, raw eggs, milk, liver, raw beef, beef bones, and salmon was Governor's regular fare.

Janet, young as she was, identified so closely with the lion that she often shared snippets of the raw meat that were the lion's basic sustenance. All through her long and vigorous life, Janet would eat raw meat and fish, sometimes ingesting small amounts of raw liver and other delicacies which most people would shudder to contemplate. In some ways her diet would resemble an Eskimo's rather than a modern-day American's. Janet went to some pains to avoid eating in this fashion in public or with those who might not understand, but she never abandoned her early affinity for a lionlike diet. Her exceptional health and stamina in later years were a testimony to how well that worked for her.

* * *

In the months before their first big trip the hotel became both home and playground to the two little girls. An alcove on their floor boasted a big bearskin rug. Janet and Kathryn discovered that when they both crawled underneath it and crept forward on all fours, they could make the furry beast come alive. They were surprised in this game one day when the elevator door opened, disgorging a woman guest who screamed bloody murder when a lumpy polar bear lurched toward her from across the hall. The girls got in trouble for that one.

Punishments were rare. George snapped their hands with his hard knuckle. Blanche withheld treats or promised activities

Life was fun at the hotel even when George was away. How many other little girls had a lion for a pet in midtown Manhattan? The furry 25-pound Christmas present became another New York attraction as she sauntered through the lobby, close on Janet's heels (leashed, of course).

* * *

Mr. Cantor, a co-owner of the Governor Clinton Hotel, became a good friend to the Hutchinsons. He was also a registered pharmacist. The hotel boasted a fully stocked pharmacy in its lobby. Mr. Cantor was solicitous and very helpful to Blanche when both girls came down with tonsillitis. He helped them find a doctor (both children ended up having tonsillectomies) and his wife administered chicken soup to the small recovering patients. The family was fortunate to find such supporters in the Cantors; the friendship and correspondence was to endure for years.

It was Mr. Cantor, as well, who helped George Hutchinson develop the final vision of the big undertaking for the Flying Family. He provided essential letters of introduction to a wide assortment of wealthy patrons and businessmen, able and hopefully willing to finance this momentous undertaking.

Slowly the shape of things to come emerged. The four Hutchinsons, plus mascot, would fly to every capital of every state and meet with every Governor, obtaining their signatures on a huge Scroll of the States. The Scroll would carry a message from the President of the United States promoting support for aviation.

Starting at the top, the Hutchinsons would meet with President Hoover in the White House and go forth from there on their mission of good will.

George Hutchinson always took the most far-reaching and ambitious view of his endeavors. Not only would this trip, in his mind, serve to reassure ordinary Americans that the skies were safe for families and children, and bring air travel into the public view in a dramatic and well publicized way: George also wanted to encourage private plane ownership. He envisioned a large array of private planes, owned by ordinary citizenry, with the potential to be converted into defensive weapons of war in times of national need.

One of George's first official supporters was Governor Ritchie of Maryland, who helped him secure an appointment with the President of the United States, Herbert Hoover, to lay out the plan and obtain endorsement at the highest level.

Things started to move, and move fast. At 12:45 PM on January 3rd, 1931, all four (five, including Governor) members of the Flying Hutchinsons met with President Hoover in the Oval Office. The President spoke kindly with each member of the family, affixed his signature to the scroll and somewhat warily shook hands with

Governor the lion. Both girls remembered the President as quiet and somewhat unimpressive.

"He didn't seem a bit afraid!" Janet Lee elbowed her sister in the ribs and whispered.

"Governor, or President Hoover?" Kathryn giggled back as the little girls stood at attention on the steps of the White House, where they posed for press photographs with the Chief Executive.

Plans and preparations for a trip of this magnitude took a good deal of time and effort. New clothes and flying togs had to be bought. Trunks were carefully packed and stored. Credit arrangements were made for refueling. Flying routs had to be planned and charted. And of course the Stinson, a sleek and shiny black and yellow monoplane, had to be checked and rechecked.

School books were one of the urgent requirements. The girls, largely home schooled, were both far enough advanced in their education to skip a grade when they resumed public school. For now, Blanche Hutchinson would rely on the Calvert System, the first home school correspondence course.

From the point of the Presidential reception on, the four members of the Flying Family became national news. The press were hot on their trail. Janet and Kathryn became possibly the most-photographed little girls of their generation. Photographs of the family from that era show a smiling and debonair George Hutchinson, usually sporting a beret. Calm and lovely, Blanche wore her fashionable cloche hat pulled low over her brow. She and her daughters wore smart flying jackets and jodphurs. Auburn-haired Kathryn met the camera's gaze with a poised and serious demeanor. Round-faced and dimpled, Janet Lee usually had her arm thrown around her leonine friend.

After the White House stop, the Hutchinsons stayed overnight at the Hotel Washington before meeting with Governor Ritchie of Maryland in Annapolis the following day. The manager made no objection to Governor bedding down with the family, so a bed was made for her in the bathtub.

The family left to go out for dinner, affixing a note to the closed bathroom door warning all and sundry: <u>Do Not Open</u>! When the unsuspecting group returned to their rooms, lights were blazing, the water was running full blast in the tub, and all the blankets and pillows were piled in a heap under one of the beds. Governor lay full length in this cozy nest, blinking up sleepily.

A maid had come to the room to leave some extra towels. Ignoring or failing to see the sign on the bathroom door, she had opened it. A startled Governor leaped out of the tub and dove under the bed, while the scared-witless maid ran out into the hall and slammed the door.

Next morning, none the worse for wear, the little lion settled down peaceably in the compartment behind the rear seat that had been built specially for her. The Flying Family were on their way. The children jostled each other at the windows to see who could spy and identify more national monuments from the air as they flew over Washington D.C. Their mission was underway.

Blanche, George, Janet Lee
& Kathryn Hutchinson
with first Flying Family plane
July 1929

Janet Lee & Kathryn reading
Richmond, Virginia
April 1929

America's Flying Family on steps of the White House
at beginning of Flying The States Flight
1931

"Cub Reporter"
Governor the Flying Lion
1931

4

UP AND AWAY

Their first official stop was on familiar ground as Governor Ritchie of Maryland became the first governor to sign the scroll. He chatted amiably with the little girls and shared their sticky candy. Crowds gathered as the Hutchinsons made their way through the winding streets of Annapolis to the airport. Soon they were winging their way to Dover, the capital of Delaware, where, due to the lack of airport or landing field, George had to touch down in a large pasture. The wheels of the plane sank into the soft earth of the muddy field. Here too, curious onlookers and press thronged, eager to see the newly commissioned Flying Family and their unusual mascot. As was to become a familiar pattern in the weeks to come, local residents gathered at the hotel where the Hutchinsons were staying to meet and chat with them.

The next day was George and Blanche's anniversary. Kathryn and Janet Lee crept out of bed early, got washed and dressed in their new flying outfits, and gleefully tiptoed through the lobby and out onto Main Street.

"Do you think we've got enough?" Janet asked anxiously.

Kathryn carefully counted the coins in the small purse again.

"$2.36. That should buy them something really nice."

As soon as they spied the florist shop, they stopped. An elderly man was sweeping the front stoop.

Janet ran up to him eagerly. "We want to buy some flowers - lots and lots of flowers. As many as you can buy for this much. Show him, Sis!" Accordingly, Kathryn opened the purse.

"It's $2.36", she said anxiously. "Is it enough? What we really wanted was red roses. Mama loves red roses."

The florist peered down into the two anxious faces.

"Well now, I think we just might be able to fix you up." His eyes twinkled under tufting white brows. "I think I know who you two are, too. You're those flying children, right?"

"How could you tell?" a surprised Kathryn asked.

"Your picture's all over the front page. Everybody in Dover knows who you are. Not to mention those strange duds," he added, eying the pair's tailored brown jodhpurs, blazers and close-fitting berets. "Now let's see what we can do about those flowers."

The proud pair returned to the hotel room bearing a ribbon-tied white box filled with three dozen American Beauty roses, nestled among ferns. The gift for their parents had cost them all of $2.00. The oohs and ahhhs were worth every penny.

* * *

On this high note, with two signatures already affixed to their scroll, the Hutchinsons set out for the office of the Governor of Delaware. He impressed the children by giving them a tour of the local jail and the infamous town whipping post. Right in the park square surrounded by government buildings stood the post where, incredibly, public whippings were still administered to those found guilty of "malicious" crimes such as wife-beating.

After a difficult takeoff from the muddy field the Hutchinsons finally gained the air in a steady, monotonous downpour. Janet Lee amused herself by cutting out paper dolls while Kathryn idly dangled a strip of paper into Governor's cage. The little lion rolled onto her back and played like a kitten.

Flying barely 200 feet above the ground because of poor visibility in the heavy rain, George followed the course of the Susquehanna River as it wound its way towards Harrrisburg, the capital of Pennsylvania. In those days most pilots still flew "by the seat of their pants." The pilot strove to maintain visual contact with the ground, relying on rivers, roads, coastlines and other navigational signposts for direction. The next step up was "dead reckoning", whereby pilots estimated direction by anticipating landmarks.

While Governor Fisher met with George Hutchinson and signed the scroll, the children were greeted and entertained by the official's grandchildren.

"Oh no! Look at Governor!" warned Janet, as the group turned to watch the cub who crouched on the carpet, entranced by a bowl of goldfish on a nearby table. Lightning-fast, Gov shot her big paw into the bowl. Withdrawing it quickly, she shook it off as the fish darted furiously around the bowl - which somehow didn't overturn. The game was highly entertaining for everyone except the fish.

Still in a downpour, the Hutchinsons resumed their flight. In Trenton, New Jersey an attempt to secure the signature of Governor Larsen met with delay. The Governor was sick with the flu and recuperating at home. Undeterred, George flew instead to busy Newark Airport, at that time the terminus point for all major airlines. They finally reached the ailing Governor at his house in Perth Amboy. Governor Larson, bedridden but hospitable, insisted on having Gov brought up from the car where she had been left with the chauffeur. The curious cub entertained herself by batting at the Governor's moving feet beneath the covers.

The little family was exhausted by the time they hit their beds in a Newark hotel. Three signatures had been added to the scroll on one very rainy day. Sleep came quickly to all, even Governor, curled up cozily in yet another porcelain bathtub.

The next morning found them underway to Hartford, Connecticut on a refreshingly clear, cold day. The trip had well and truly begun.

* * *

The interior of the monoplane seated five people, two in front and three in the back. Behind the back seat was the baggage compartment and Governor's roomy cage. Wide windows that rolled up and down manually lined the sides of the cabin. The children could stretch out for naps on the wide, single back seat. Six feet of head space made moving around easy. Altogether, it was a comfortable home in the air.

The bond between Governor and Janet grew stronger day by day. Surreptitiously, she sneaked tidbits to her pet between meals. Each night they wrestled together as though they were a pair of puppies. It was a strange sight to see Governor taking a breather, sitting on top of Janet as she lay prone on the floor. Gov took to trailing along after Janet everywhere she went, even without a leash.

A routine was established from the beginning of the trip. In the mornings, George lowered the cabin window regardless of the icy air and led the family in deep-breathing exercises. Then came ten minutes of situps, with the females stretching out on the seats and George guiding the plane with only his feet on the rudder.

Next came singing. George, a former choirboy, had a beautiful voice, although Blanche couldn't carry a tune in a bucket. Everything from *The Star Spangled Banner* to *Polly Wolly Doodle* became part of their repertoire, followed by routine singing scales.

Then it was time for the "Classroom in the Skies". The rear wall of the cabin, directly over the baggage compartment, held rolls of large pulldown maps and a five-foot blackboard. These were unfurled and Blanche took over as teacher for the next hour or two. She read to the girls from history and geography books, relating the material to the territory over which they were flying. Then the girls would be quizzed by their father on what they had learned. To end the morning session, Janet Lee and Kathryn drew on the blackboard from memory the route they had just followed and the most important landmarks.

Over the course of their trips, both Kathryn and Janet became truly knowledgeable in geography and history. Later classrooms became fairly mundane to the girls after their vital learning experiences in their Classroom in the Skies.

As their trip progressed up the Eastern Seaboard, it became like a game to the Hutchinsons to see how many states they could visit in the shortest amount of time. Connecticut, Rhode Island, and Massachusetts were all covered in a single day.

Flying from Boston to Concord, New Hampshire, the plane was engulfed in a heavy snowstorm. Lessons were put aside as Blanche concentrated on the maps and her husband busied himself with navigation. The swirling snow muffled the sound of the motor and magically softened the world around them as the girls watched the huge flakes splat against the windows.

Landing at the snow-mantled airport was another matter. The plane plowed heavily into the soft snow and became firmly stuck. An airport tractor towed them to the hangar.

Their African transplant gloried in the wonders of a snow-filled world. The cub was cautious at first, picking up her big paws and shaking them fastidiously. When she saw the girls plunge gleefully into the strange white stuff, though, she got the idea. Soon she was racing through the snow and scooping it up in her mouth.

The family accepted an invitation to go tobogganing on a nearby steep hill where lovers of sledding and skiing had gathered. The schools were closed, and the slopes were packed with children. Governor, of course, trailed along.

Soon the word spread and whole families were showing up to see the lion cub. After a good deal of pleading, the girls persuaded their mother to let them take the lion down with them and six other children on the large flat toboggan.

Reluctant at first, Governor was persuaded to settle in between Janet's legs. Both girls held on tight to her leash. Then someone, without warning, gave the toboggan a huge push and away it raced downhill, every other sled following it. About halfway down the slope the speed, the wind in her face and the screams of the children around her were too much for Gov. Roaring mightily, she began to scramble back toward the rear of the toboggan, vaulting right over Kathryn and Janet. They were sent flying off the sled, tumbling over and over in the snow. The other children abandoned ship and rolled off too, preferring to take their chances with the hill rather than an upset lion. Only Governor remained on board, sitting up straight as a board and squinting into the wind, as the toboggan coasted to a stop at the bottom of the hill.

The Hutchinsons tried skiing for the first time on the lawn of the Governor's mansion. Janet, inexperienced but fiercely determined, kept struggling to master her skiis for so long that, oblivious to time, she wet her pants. But all in all, New Hampshire was great fun. The perils of winter travel in a monoplane had yet to surface.

* * *

Snowstorms continued to plague the Flying Family as they flew north to Augusta, the capital of Maine. Here, there was no real airport or hangars, just a parade field covered with the deepest snow they had yet encountered. It was a real possibility, George warned, that the plane could actually capsize in the drifts. Sure enough, when he landed in the deep snow the nose went down and the tail came up. For a moment it felt and looked like the plane would somersault. Onlookers held their breaths in anticipation of a real catastrophe. Quickly, George unsnapped his safety belt and jumped over his seat into the back of the cabin. His 160 pounds threw the balance of the plane in the opposite direction and it settled back down into normal landing position with no harm done.

"That was fun! Let's do it again, Daddy," crowed Janet Lee.

Riding in style in a horse-pulled sleigh to the home of Governor Gardiner of Maine was a novel experience for all, not least for the lion cub and the horse.

Getting out of the state proved even more difficult. George decided to make a takeoff from a nearby highway, the only cleared ground available. This was a tricky proposition as it required straight-as-an-arrow taxiing to avoid power lines and buildings lining the roadway. Once again the Hutchinsons gained the skies with impunity and were off to Albany.

This time, their luck didn't hold. Their projected course to Albany led them straight into a blizzard. Reluctantly, George turned back to the little airport in Portland, Maine. The Hutchinsons just made it to the hangar before the storm broke in all its fury.

During the night the snow turned to heavy rain. As they resumed course toward Albany the next morning, the Stinson's motor began to sputter just over Pittsfield, Massachusetts. Most likely water had gotten into the carburetor or gas line. Over the town the engine shut down entirely.

"Quick, everybody! Fasten your safety belts—we're going down!" George shouted.

Deep snows remained on the ground where the plane slipped down onto a small rolling field. This time, when the wheels hit, the plane nosed straight up. The propeller struck the ground, badly bending the blades. The tail was sticking straight up in the air. People had come running from all directions when they saw the impact. Now, they watched the four Hutchinsons jump adroitly down from the jacknifed plane onto the soft snow on the ground.

Delays for repairs kept the Flying Family grounded for days.

5

NORTH BY NORTHWEST

Why, one may ask, did an intelligent man and accomplished pilot such as George Hutchinson choose to head north in January, flying his family into potentially dangerous weather conditions? Why not head south or southwest in the coldest months?

If there was indeed a pragmatic reason, no one knows what it was. He may have favored a northern start simply so that, in covering several smaller New England states quickly, he could amass the greatest number of signatures in the shortest amount of time. Certainly it was well known that George enjoyed a challenge. His belief in his own infallibility reinforced by the absolute confidence of his children and the unwavering loyalty of his wife perhaps contributed to a disregard for ordinary cautions that might affect lesser mortals. Possibly, it did not even occur to him to be concerned over winter weather one way or the other. Picking the more prudent course of leaving the northern states until Spring, or even delaying the whole trip a few months, didn't jibe with George's supreme self-confidence. The more obstacles, the better he liked it.

News of the nose-dive landing flashed around the nation via radio and newspaper reports. The name "Flying Family" was fast becoming a household word. A deluge of telegrams and letters poured in. Some were supportive, some decidedly critical. Why, critics asked, was George Hutchinson putting his children in jeopardy?

Franklin Delano Roosevelt was still Governor of New York at this time, residing in the state capital in Albany. Meeting the great man for the first time, The Hutchinsons were greatly impressed with his sincerity and warmth.

More extensive repairs and a tour of the Stinson Airplane Factory in Wayne, Michigan, where the family plane had been built, delayed the trip still further.

* * *

Skies cleared as the flight resumed through the northern central states and proceeded northwest through North and South Dakota. In Pierre, South Dakota, they were given the same rooms occupied by Charles Lindbergh during his prior visit to the city. The girls were thrilled.

Blanche taught Kathryn and Janet about Native American tribes as they flew over the broad plains of Wyoming. Within 50 miles of the capital, George was alerted by a column of smoke rising from the ground below. A small ranch house was engulfed in flames. Three men were fighting the fire frantically. Alarmed, the Hutchinsons spied the body of a woman lying on the ground a short distance from the burning house.

Picking out a smooth stretch of barren ground, George landed and taxied as close to the house as possible. Blanche and George rushed to the side of the prone figure. Fortunately, the woman wasn't burned as badly as they had feared. After ministering to her from the plane's first aid kit, they realized she would be fine. George took with him a hastily scrawled message for help wrapped around a chunk of wood and took off again. Unable to land near the closest town because of the rough terrain, they dropped their message from the window of their circling plane onto the steps of the general store. George waited to see it picked up, dipped his wings in farewell and headed for the state capital of Cheyenne.

* * *

Flying across the majestic mountain range that separated the prairie from the high peaks of the Rockies, the Hutchinsons finally approached Denver, Colorado.

In their well-appointed Denver hotel, Kathryn had a frightening encounter. Poking around the hotel alone, she spied a lavish parlor with a grand piano in the middle of it. Kathryn dearly loved to play the piano. Checking to make sure she was alone, began banging away happily. Suddenly the door slid open. An elegantly dressed elderly man entered and, uninvited, joined her on the bench.

Trained to be courteous around grownups, Kathryn didn't know how to react as the overly-friendly man began to play along with her. Surreptitiously his hand crept onto her knee and and began inching up her dress. Badly frightened, Kathryn slipped eel-like from the bench and ran out the door. Bewildered and embarrassed, she never told anyone about her close call.

The next day, in response to Janet Lee's begging, the family visited a big cattle ranch. The girls were excited at the prospect of riding Pinto ponies over the range like real cowgirls. They marveled at the intricately carved Western saddles, having known only the Eastern style back home. Each "tenderfoot" had minor mishaps on horseback on that trail ride, but only Janet managed to get thrown by her mount, rolling under the horse's hoofs. She picked herself up and remounted immediately, eager to ride some more. The small equestrienne seemed to be fine.

That night at the hotel, it was time for some much-needed grooming for Governor. Her sharp claws needed to be clipped. The whole family had to hold her down, while she protested so loudly that alarmed guests reported the uproar to the management. The manager and two bellboys rushed to the Hutchinsons' room, but ended up helping everyone hold the lion down so the task could be completed.

When Blanche lifted Janet Lee into the bathtub that night, she was shocked to see scrapes and dried blood covering the little girl's ribs on the left side.

"What in the world happened to you?"

"I guess maybe the horse kicked me when I rolled off him. But he didn't mean to!"

"But why didn't you say anything, honey?"

Janet ducked her head. "I was afraid you wouldn't let me ride a horse again. And I just love horses! I'd never tell on a horse, no matter what he did."

* * *

Next morning, the family enjoyed perfect flying weather from Denver to Rock Springs, Colorado. The Stinson soared at 12,000 feet above some of the highest peaks in the Rockies. The children gazed down in awe from this great height at the shining white peaks below. Rivers and roads snaked like ribbons between passes in the great ranges. Lakes and tiny towns lay all but buried in the deep valleys. On a jutting crag a mountain goat stood silhouetted against the sun.

Landing to refuel in Rock Springs, George was warned by the pilot of a mail plane at the tiny airport that heavy fog and snow lay ahead on the approach to Salt Lake City. George decided to take the chance, as he usually did. As they approached the Mormon city the plane was engulfed in deep mist with zero visibility. When they were only some 300 feet above the ground the golden dome of the great Mormon Temple appeared like magic out of the fog and they were able to land at the airport shortly beyond it. Due to flying at such high altitude for an extended period, no one in the family could hear well for a few hours. Pressurized cabins, of course, were unheard of at that time.

After securing the signature of the Governor of Utah, the Hutchinsons made a friend of a professional wrestler and cousin of the Governor in the lobby of the Utah Hotel. The wrestler presented the family with tickets for a local wrestling match scheduled for that night.

The whole family, lion included, attended. Governor, curled up on the ringside seat beside Janet, went unnoticed or was perhaps mistaken for a large dog in all the excitement. Just before the main event, while the fighters were removing their robes in their respective corners, the announcer introduced the Flying Family and gave a short speech about their exploits. The Hutchinsons were invited onstage.

As George began to climb through the ropes, he was almost tripped up by Governor, who hurtled past him and began scampering in circles all around the brightly lit ring.

"What the - !" "What *is* that thing!" "It's some kind of damn wild critter!"

The hulking wrestlers jumped over, under and through the ropes in their hurry to escape the marauding beast. Calmly, Janet Lee stepped into the ring, picked up her pet in her arms and climbed back through the ropes with her squirming load of excited lion. The audience roared with laughter. Governor roared in victory.

* * *

Idaho and Montana followed in fast succession but in worsening weather conditions. The heat from the engine exhaust kept the cabin of the plane comfortably warm, or at least warmer than the unheated hangars and airports the Hutchinsons routinely encountered in that sparsely settled part of the country. As they made their way west toward Olympia, Washington, the fog thickened. Only the very tops of the highest mountains were visible, rising like islands from the engulfing grayness.

Blanche began to worry. It wouldn't be possible to make a safe landing in this mountainous terrain if they had an emergency.

George tried to reassure his family.

"Just look at all the sunshine and blue skies up here where we are! Come on, Blanche, smile. Show me your pretty teeth!"

The teasing drew little more than a tight-lipped grimace from his wife as the plane was buffeted about more strongly by the escalating winds.

George didn't realize that his plane was caught up in wind drift that was carrying it seriously off course to the north. Worse, the radio had stopped working; the storm had snapped the aerial right off.

The end of the fourth hour in the air was rapidly approaching, and even George began to really worry. The fuel tank had only a four-hour capacity. Somehow, he would have to find a hole in this impenetrable mass and land. At 11,000 feet he headed into the thick whiteness.

"I can't see anything, Daddy" Kathryn said anxiously.

"Just old white fog," added Janet.

"Well, why don't we make a game out of it? Kathryn, you take the right window. Janet Lee, you look out the left. Mother takes the front right and I'll watch my side. Look just as hard as you can. The first one that finds a hole in the fog - a glimpse of trees, mountain, ground, anything - yell out, and you'll win the prize."

"What prize?" Excitement started to build.

"Anything you want, costing up to $10, when we land."

A tense silence prevailed as they descended as slowly and cautiously as George could manage without stalling out. Minutes passed. Still no break in the fog. George prayed that his gas would hold out, that they would not emerge from blindness only to crash into some great wall of rock.

Four thousand feet. Three thousand. Two.

Suddenly, "Trees, Daddy! It's trees!" Janet Lee shouted.

The altimeter arrow pointed at 500 feet.

A hairpin bank to the side, and the Stinson plowed through the rift in the mist.

"Buckle your safety belts, everybody! Quick!"

A small open space, surrounded by heavy fir forest, emerged below. With only 200 feet remaining between him and the ground, George banked steeply and skimmed the tops of the trees as he leveled off and touched down into the snow.

Here, the famous Hutchinson luck finally gave out. The plane plowed heavily into the five-foot-deep snow, flipped over like a turtle toppling in slow motion, and lay on its back with who knew what damage done to the wrecked undercarriage, now pointing toward the heavens.

6

CRASH COURSE

The silence following the crash seemed to echo in Janet Lee's stunned ears for an eternity. The first thing she became aware of was the rasping of a warm wet tongue licking her scalp, where a huge bump was rising rapidly from the impact of her head with the edge of Governor's cage. Then she noticed she was hanging upside-down from what was now the ceiling of the cabin, still strapped in by her seat belt. Governor had been thrown free of her cage and was now attending to Janet's wound.

George quickly unfastened his own belt and helped the rest of the family to right themselves. Oil from the fuel tank was slowly seeping through the firewall into the cabin. It was time to abandon ship.

Finally, the little family found themselves sitting shakily on the edge of the wing. Everybody checked everybody else out for injuries. Aside from Janet's nicked head, the children had sustained only bruises and sore muscles. Blanche too seemed basically uninjured. George had a blackened right eye and cut knee, poking through his torn jodhpurs

Governor limped on her left forepaw. Still, she continued to play nurse to Janet Lee, pushing her furry body as close as she could to the little girl's side and licking her face and hands and head with her rough tongue.

Walking slowly around the Stinson to assess the damages, George's spirits fell. The motor had been completely torn from its mounting and the undercarriage was badly bent on one side and severed on the other. Both wings were warped and twisted, and the propeller was bent. Only the cabin, fuselage, and tail seemed intact and undamaged.

Where were they? The fog still carpeted the dense forest around them as far as the eye could see. A brief survey in each direction revealed only more trees.

Three long hours passed with no sight or sound of humanity. Only the lone cries of birds rang out in the shrouded stillness.

"What's that!" Kathryn and Janet both jumped as a mournful keening sounded from the distance. It was the howl of a timber wolf.

"It sounds just like a dog. Do you think he's lost?" Janet Lee, ever the animal lover, was worried.

"No, honey. He's just singing. Wolves love to sing. Maybe he's calling his mate."

"What's a mate, Mother?" asked Kathryn.

"Never mind that for now. We need to help your father."

Each family member gathered their belongings from the upside-down cabin, piling them on the wing. While George struggled to remove the damaged propeller and motor, Blanche and the girls kept in motion handing him tools. Anything to keep warm. Then they went in search of firewood, Governor limping along behind them. All were beginning to feel the damp cold penetrate their bones.

Pickings in the deep snow were slim, and the little group returned to the plane with only a small armful of wood.

"This is wonderful," said George. "Just wait and see how warm we'll be in a few minutes." Draining the last drops of gasoline from the tanks to help start the blaze, he discovered there had been exactly one quart left. Only one more minute of flying time. Maybe their luck had held, after all.

The family huddled gratefully around the warm blaze, toasting hands that were stiff with cold and toes that had turned into ice cubes, trying to ignore the growing dusk that was nearly upon them.

* * *

"Whoa! Whoa there!"

Suddenly the silence was broken by a man's deep shout from the woods nearby. Electrified, George bounded through the snow in the direction of the sound. A sledge piled with logs, pulled by a huge dray horse, emerged from the woods. A very tall man held and flicked the reins.

The woodsman stopped short in amazement at the sight of the begraggled Hutchinsons around the dwindling fire and the hulk of the wrecked plane. Hurried explanations and introductions were made, and the children soon found themselves perched atop the log sleigh with the family luggage while the adults walked alongside. They were heading for the five-mile distant town of Kitchener, British Columbia. The northerly drift had landed them severely off course - in Canada rather than Washington.

Mr. Redmile, their rescuer, kept the family entertained on the cold trek with information about Kitchener, whose population at last count had reached 50. George was interested in hearing about the local logging industry and camps. The girls were intrigued by his tales of his ranch and brand-new litter of 13 newborn pigs.

The proprietor of Kitchener's only hotel welcomed the shipwrecked family and their strange pet. The owner of the general store, proud possessor of the town's only telephone, sent news of the crash to the nearest newspaper. Word traveled fast in that isolated community. That very night, townsfolk flocked to the hotel to meet the strange guests.

But George had much to attend to before he could rest for the night. The town doctor looked over everyone to be sure all was well. Except for bruises, everyone was pronounced to be in perfect health. Canadian Customs had to be notified and inspection passed. Lloyds of London, insurers for the plane, had to be notified, as did Mr. Cantor and worried relatives. Most daunting, plans had to be made to dismantle the Stinson and ship it off to an airport for repairs.

As bedtime finally approached, the manager balked at the idea of a lion, however tame, inhabiting one of his guest rooms. By this time Governor regularly bunked at the foot of Janet's bed. But the manager held firm. Governor would have to be tethered in the lobby. The little lion hated sleeping where it was cold. Only a wood stove heated the entire building, and it was decided to tie Governor to a chair beside the stove. A weary Flying Family trudged up to their adjoining rooms on the second floor.

As George bent down to tuck Janet in for the night, she whispered in his ear.

"You probably haven't had time to think about it yet, but remember, Daddy, I did win the ten-dollar prize."

* * *

The exhausted family fell into their beds at last, anticipating a well-earned rest. One by one the villagers left the warm hotel lobby for their homes and soon the lights were out, only the heat from the stove casting a warm glow in the darkness.

The great Canadian woods grew colder, the snow-capped trees standing like sentinels in the vast wilderness. All the humans in the hotel slept deeply, and for a while, so did the tired little lion, curled as close to the stove as she could get without singeing her fur.

The silent hours passed. The fire grew low and gradually died out. As frigid air slipped under the door sill, Governor grew cold and woke. Finding herself chilled and alone, she was no longer content to remain tethered. Where were her people? With fierce concentration she turned to tearing and biting the leather leash that held her. Finally, it broke, and following her nose she made a beeline for the stair.

Governor found the door to the children's room closed. Heaving a great sigh she lay down by Janet's door. Soon the cold draft drove her to pad down the hall once more, searching for a warm nest. Surely this hotel had better accommodations than this!

Finally at the end of the hall she spied a wide-open door. Silently she entered, big paws making no sound on the rough wooden floor. The bed looked warm and inviting. And after all, hadn't she slept on one many times before? She crawled up on the soft featherbed and burrowed under the covers. Now this was more like it.

All of a sudden a rattling and scratching sounded at the front door as someone fumbled with a key. Finally the door swung open and an elderly man, somewhat the worse for wear from drink, wove his way across the lobby and up the stairs. Donning a white nightshirt and warm nightcap with a long tail, he gratefully collapsed into his soft featherbed.

The fog of alcohol lifted just enough for the awareness of something not quite right to infiltrate the brain of the sleeper. He turned over and opened his eyes.

Governor opened hers, too. In the dimness the lion's eyes glowed, big as half dollars. She opened her mouth in a wide grin, showing all her white teeth, and let out a friendly roar.

A piercing scream echoed throughout the hotel. .

"Help! Murder! Help! There's an elephant in my bed!" The man's cries roused the whole house. He leaped out of bed, ran down the stairs and out the front door and out into the snow, clad only in his nightshirt. He did not return that night.

When the general uproar settled down, Janet Lee took Governor into her own warm bed for the night.

"Just think how hurt her feelings must be, Mama," she explained when her mother shook her head over the arrangement. "Governor never met anybody that didn't like her before. That man scared her pretty bad."

* * *

When George surveyed the site of the crash the next day, he realized just how lucky they had been to land in the only open field for miles around. The sadly crumpled plane was surrounded on all sides by giant trees and snow-covered mountains. Redmile and a local crew would spend the day dismantling the crippled plane and loading the parts on a box sled. Then, on to the train station where the plane parts would be shipped to the airport in Spokane for repairs.

Meanwhile, back at the hotel, Blanche stayed in bed nursing her bruises and sore muscles while the more-resilient youngsters took to the snowy hills nearby with a large dishpan they had borrowed from the cook to use as a sled. Governor bounded beside them as they flew down the slope and trudged back up. She was the center of attention for the crowd of local children who looked on in amazement as the little lion capered through the snowy drifts.

That night a tired and hungry contingent of Hutchinsons gathered once more around the table at dinner. The day in bed had done Blanche a world of good.

The children rattled on about the day's discoveries.

"We saw this cave, didn't we Sis, where great big bears live. Except they're asleep now. They sleep for weeks and weeks," Janet eagerly explained.

"The word is *hibernate*" interrupted the more erudite Kathryn. "Bears hibernate, all winter long. They eat and eat and then they live off their fat."

"Anyway, we didn't make any noise." Janet continued. "We didn't want to wake them up so we tiptoed. We tried to get Governor to go in by herself but she wouldn't. An Africa lion shouldn't be afraid of an old bear, should they? Say, Mr. Redmile, can we see your baby pigs tomorrow?"

The long trestle table seated fourteen people that night, all of them interested in meeting the Flying Family and hearing about their adventures. More locals dropped in later, gathering around the big stove in the lobby, to see the lion and hear the news. Unlike the previous evening, tonight's sleep was peaceful for all, with Governor curled snugly at the foot of Janet's bed. Around midnight, George awoke briefly when the freight cars carrying the disassembled plane to Spokane got underway, banging and clacking down the track.

The morning brought the promised visit to the Redmile ranch. The night had been so cold that Mr. Redmile had brought the piglets inside the cabin to keep them warm by the stove. The girls were horrified to be told that the old sow, upon their return, had killed four of them - evidently acting on instinct to destroy her young rather than have them taken away. Now the nine remaining piglets would have to be kept inside until they grew large enough to be safe on their own.

Janet and Kathryn, disturbed over the thought of a mother destroying her own babies, were diverted by Rex the big black German Shepherd, who played gently as a lamb with a tiny white kitten who ran up and down his back. Once again, laughter prevailed.

* * *

The girls hated to leave their new friends behind as their train pulled out of the tiny whistlestop for the six-hour run to Spokane. The awesome beauty of the high Canadian Rockies surrounded them. Everyone on the train was eventually introduced to the

Hutchinsons, persuading the family to share the story of their adventures over and over. Finally, they reached the homelike Davenport Hotel in Spokane, only to be faced with a seemingly endless delay.

New parts and wings for the Stinson had to be ordered. The propeller had to be returned (yet again) to the factory in Pittsburgh. Insurance, finances, and shipping delays interfered. Days turned into weeks and the weeks piled up. The children grew restless, their parents grew impatient, and Governor the lion just plain grew.

Partly to fill the empty time, the family accepted invitations to speak before many local organizations and schools. Kathryn and Janet, young as they were, were becoming accustomed to the limelight and to telling about their adventures. Governor, of course, was always the center attraction.

Governor was growing more accomplished in the matter of tricks, too. Like a well-trained dog, she could sit up, roll over, and fetch the newspaper. She could chase and retrieve a ball, play hide-and-seek, and of course box and wrestle with the children. The sight of a lion cub performing all these antics was so funny and entertaining, the parade of people craving to see her seemed endless. The Hutchinsons stayed so long in the Spokane hotel that in the end the management built a kennel for the cub on the roof.

But Governor's favorite spot in the hotel was the big fountain in the lobby with its water lilies and enticing goldfish. She made a beeline for it every time they passed through, crouching patiently beside the pool until some unsuspecting fish was unwise enough to show itself. Quick as lightning, she swiped at the water with her big wobbly paw, but never succeeded in actually spearing one of the maddening creatures. Yawning with pretended ennui, she shook off her wet paws and stalked haughtily away, leaving wet tracks across the carpet. She never tired of the game.

To Governor, the world was full of friendly people who wanted to make much of her. One day in the lobby she leaped up on the lap of a woman reading a newspaper. Frozen, the lady found herself staring into big round yellow eyes and a gaping, fang-filled smile. She was absolutely speechless. Taking silence for consent, Gov proceeded to lick her face with her rough tongue.

Hungry for entertainment, the family took in a local high school basketball game. The team mascot was a cougar, and the boys begged to borrow Governor as a live stand-in for the night's game. All went well until the ball accidentally bounced off the cub's head. Gov grabbed it by the lacing and a free-for-all ensued as the team took out after their ball. Up and over chairs and across the court the chase raged. Half afraid, the boys finally cornered the lion who turned to growl back at them ferociously, one paw on the ball. One brave player approached the snarling feline slowly and carefully. Governor crouched for the spring. The boy grabbed for her collar, but she made a lightening-quick turn and he landed face-down on the floor.

Shouts rose from the crowd. "Get a rope!" "Get a gun!"

But the cub immediately jumped on the boy's back and proceeded to give the back of his ears a thorough tongue-bath. Tensions eased as laughter rolled through the gymnasium.

* * *

Finally, after five weeks of waiting, the Stinson was ready to fly. With new wings, dowling, undercarriage and other parts replaced she looked as good as new. Scores of their new friends were at the airport to see them off as the Flying Family took triumphantly to the air once again, heading due west this time towards Olympia.

Unbelievably, mechanical trouble reared its head yet again while crossing the awesomely beautiful Columbia River. The engine sputtered and died with no warning. Spying an island in the middle of the river, George maneuvered a safe landing. Thankfully, this time there was no structural damage to plane or occupants.

Stranded again, this time on a small strip of land only 1500 feet long, the Hutchinsons took stock. The little island was covered with weeds and small trees, smack in the middle of the mighty and swift-flowing Columbia. The high banks of the Columbia Gorge rose on either side.

George hated to radio for help when if there was a chance he would be able to fix the trouble himself. He spent hours checking out every part of the motor, finally getting it to turn over, only to sputter and die once more. Meanwhile, Blanche and the children explored the island.

"There's nothing here, Daddy," said Kathryn. "No boats or rafts or even logs. And the water's way too swift for swimming."

"Oh boy, we're marooned!" said Janet. "Just like those Swiss Family Robinsons!"

"I think we'll have to spend the night in the plane, then," sighed Blanche.

Long into the night George worked on the motor, the girls holding flashlights and handing him tools. Dinner was candy bars, oranges, and a few apples.

Blanche and the girls were still sleeping in their improvised sleeping nests inside the plane the next morning when George crept out to do a little fishing for breakfast. Finding a safety pin to bend into a fishook, he attached it to a ball of twine and tied it to a long branch. Before long he had pulled five fish from the swift currents of the Columbia. Returning to the plane he gathered driftwood, twigs, and branches for a fire. Soon the tantalizing aroma of freshly grilling fish wafted into the open window of the plane, awakening the sleepyheads who were thrilled with their surprise breakfast.

This morning's inspection of the motor revealed the problem: a small amount of steel shavings, carelessly dropped in during the repair work, had wedged in the gas line leading from the left wing tank, blocking the free flow of gas into the carburetor. Once the right tank had drained empty, the block in the left line meant no more gas to the engine, and thus the forced landing. After draining the tank and patiently fishing around with a screwdriver and wire, George managed to withdraw the shavings.

After cold baths in the river and a change of clothes, everyone felt refreshed and ready for the next big challenge, getting the plane off the island. It was a tricky business, with the island's rough terrain and extremely short space for taxiing, but this time fortune smiled and soon once more they were on their way.

7

THE WILD WILD WEST

George was anxious to make up for lost time. Rapidly over the next few days he obtained signatures from the governors of Washington, Oregon, Nevada, and California. Each day was a grindstone of flying, eating, sleeping, and getting from here to there as quickly as possible.

The Hutchinsons couldn't resist an extended stopover in Los Angeles, however, and the chance to become true tourists for a few days. The Flying Family visited all the major studios, and were welcomed onto the sets of several movies to watch the actual filming. They were able to meet many famous stars: Gary Cooper (with whom Janet would star on radio dramas a few years later), Wallace Beery, Jackie Cooper. The four swam in the Pacific and toured Venice, Ocean Park, and Long Beach. They picked oranges straight off the trees and visited all the typical tourist haunts such as Sunset Boulevard and Groman's Chinese Theater. It was hard to bring themselves to leave California but there were still seventeen more signatures to obtain, and seventeen more states to visit, before their mission was complete.

The next scheduled stop was Phoenix, Arizona, but since they found themselves so close to the San Diego Naval Air Base where the great Pacific Fleet was anchored, they decided to fly by way of San Diego. Following the coast north on a glorious day, the distance was covered in less than an hour. George set down on North Island, the government air base. Strangely, the airport seemed almost deserted. As they taxied to the hangar, a man who had been working on a plane loped over to intercept them.

"I think he wants to talk to us, Daddy" said Janet Lee. "And I think he has a camera with him."

"Probably another reporter," murmured Blanche.

"Nope, that's a motion picture camera he's toting" said George, leaping down to the tarmac and advancing to meet the stranger.

The man held out his hand. "Hi folks, I'm Lieutenant Hamilton," he said with a grin. Introductions all around followed.

"I'm stationed here at the base. The plane I was working on over there went haywire, and every other available plane left the base a little while ago, to join today's air and sea maneuvers. I was supposed to document it from the air, but..." He glanced around in frustration. "Would you be able to spare your plane, and some time, to fly me out over the Fleet?"

George agreed readily. The eager Lieutenant ran to get the rest of his equipment, while Janet and Kathryn begged to go too. Thus the family Stinson became the official photographic plane for the Pacific Fleet for one day. As they flew out over the sparkling Pacific the two men worked from official position forms and maps to locate the fleet, and soon spied smoke from the huge concentration of ships on the horizon ahead.

Janet Lee and Kathryn became possibly the only children to view a great navy in warfare simulation games. It all seemed frighteningly real. Cannons belched fire and smoke, shells whistled and exploded, and the wall of smoke laid down by airplanes blanketed the entire fleet. Hundreds of planes in tight formation took off from the flight decks of carriers and dropped bomb after bomb on the wrecked ships that were used as targets.

Hamilton swung from one side of the plane to the other, capturing dramatic views in all directions. George maneuvered the plane adroitly to keep it in proper range and position.

From high altitudes of 10,000 feet the ships below looked like toys, but from the lower altitudes of 100 feet or so the noise, the impact of targets exploding, the swirling smoke and planes darting like swarms of angry hornets through the incongruously blue skies above looked, and felt, like real warfare.

Blanche shuddered slightly and reached for the her daughters' hands.

Suddenly, just as the Lieutenant was shooting the last of his 6000 feet of film, the Stinson was totally surrounded by a formation of nine naval planes.

"Look, Daddy - they're saying hello!" exclaimed Janet Lee, as the family smiled and waved at the pilots. But the soldiers didn't smile back. Beneath their dark goggles, expressions looked grim. Each pilot signaled by pointing downward. They wanted the Hutchinsons to land, and land now.

Hamilton couldn't get anyone to recognize him out the cockpit windows. Inexorably, their plane was being forced lower and lower. Tension was high. Now they were being drawn into this game of war for real.

"Where the blazes do they expect me to land," shouted George, "on the water?"

"No," said the Lieutenant, "on one of the airplane carriers."

The formation nudged the plane down in a straight line toward the huge airline carrier bringing up the rear in the line of ships. George touched down on the long deck in a perfect landing. As soon as they rolled to a stop, two sailors yanked open the cabin doors and an officer marched up.

"Who's the owner of this plane? Whose camera equipment is this?" he barked.

George Hutchinson and Lieutenant Hamilton were led off to meet with the brass, leaving a nervous family waiting in the plane. The sailors saw to it that they stayed put.

Janet turned a worried face up to her mother. "Is Daddy arrested?"

"They can't arrest Daddy," Kathryn said indignantly. "He was only helping out!"

Blanche reassured the girls that everything would be explained in short order. And sure enough, soon George and the Lieutenant came strolling back, surrounded by a contingent of laughing and chatting officers. The Hutchinsons were thanked instead of incarcerated, and given an escorted tour of the huge ship

Thus for the first and last time the Flying Family took off from the deck of an airplane carrier. As they gained altitude George dipped his wings in farewell and flew straight eastward, toward Arizona and their thirty-second capital. The signature of Governor Hunt was quickly obtained, and they headed for Santa Fe, New Mexico.

* * *

During this leg of the trip Blanche read to the children about the Cliff Dwellers, the ancient peoples who had once lived in this part of the country. The girls were fascinated, marveling at how whole families could possibly live in caves carved into the sides of steep cliffs.

As they flew past high mountain ranges, George noticed excavated dwellings pocking the cliff walls. He decided to surprise the family by landing nearby and investigating.

The plane rolled to a stop at the very edge of a barren plateau, overlooking a deep ravine.

The children were intrigued. "Are we in Santa Fe already?" asked Janet Lee.

George laughed. "Not yet, but we are at the homes of the Cliff Dwellers - just like Mother was telling you about. Care to investigate?"

Soon the four were carefully picking their way down the face of the ravine, the cub following on her leash. At the bottom they crossed to the other side and stood peering up at the ancient dwellings. There seemed to be no way up to the high entrances, but further down the valley they found rough steps cut into the sandstone. Janet Lee scurried ahead to be the first to reach the lowest, most accessible level.

The small cavelike dwellings dug into the side of the sandstone cliffs reminded the children of modern apartments stacked one atop each other. They receded toward the top of the cliff high above.

Gradually a sense of timelessness enveloped the Hutchinsons. Paintings and strange symbols decorated the smoke-darkened stone walls of the rooms. Rotted grass mats covered some of the windows and doorways. Their voices grew hushed. This was truly history made real and vital.

"Wait! What's that?" whispered Kathryn. A rattling, knocking sound came from the adjoining room.

"Maybe we should get out of here. Something's moving!" cried Janet Lee. "Maybe it's a ghost!"

"Stay here with your mother. I'm going to find out what that noise is," said their father, lighting a match and moving through the low doorway leading into the next darkened chamber. He had to stoop to avoid hitting his head on the arch.

Again came the rattling, this time even louder, and with it a scuffling sound.

Janet Lee clutched her sister's hand and squeezed hard. "Ouch!" protested Kathryn.

"Shhh, girls. George, what is it?" called Blanche nervously.

George struck another match as the cautious trio peered through the passageway behind him. He laughed.

"Here's your ghost, girls! It's just our big pussy cat, gnawing on a bone she found."

The lit match illuminated the cat's golden eyes and the white flash of a bone clutched firmly between her big paws.

George knelt to examine the pile of bones. A chill struck his marrow.

"My God, these are human bones. Skulls and all," he whispered.

Kathryn, Janet Lee, and Blanche all scrambled back out the narrow doorway. Making it back to the plane in record time, the Flying Family lost no time in leaving this place of bones and ancient mysteries far behind.

Only many years later was controversy to emerge publicly about whether the ancient cliff dwellers of that region such as the Anasazi had indeed practiced cannibalism.

Were the spirits of the departed ancients disturbed at having their centuries-old secrets unearthed? Perhaps, for as the family took off once more toward Santa Fe they were to come face to face with a new and potentially deadly phenomenon of the Southwest.

* * *

It came on fast. Out of a clear blue sky, sudden clouds gathered and thickened around them. Heavy rain began to buffet the little plane. A strange dun-colored vapor seemed to rise from the bed of the desert floor like a solid wall.

"What is it, Daddy?" exclaimed Janet Lee, straining to see through the ugly murk that seemed to have instantaneously swallowed the plane whole.

"Yuck! Close the window!" Kathryn struggled to shut the open window through which gritty dust swirled, getting into their eyes and noses and throats.

But the choking dust continued to pour into the plane from the engine and fuselage. Everyone began to cough.

George realized that they had already penetrated miles into this storm. He tried desperately to bank hard and outrun it, but the dust made breathing too difficult. They had to land and land now if they were to survive.

Down, straight down they dove, leveling off at the last moment to land on the sage-covered prairie. The cabin was so full of dust it seemed better to take their chances outside, so they jumped from the plane and sheltered underneath the wings, lying flat on the ground.

"Girls! Cover your heads with these!" Blanche and George pulled coats and towels from the plane and everyone huddled in silence, as the minutes seemed to stretch out into hours, with no sound but the howling winds of the prairie around them.

Night was falling. A queer, impenetrable darkness pressed down upon them. The clouds above opened once more and huge, yellow, muddy raindrops began to fall.

Thankfully, the rain had cleared the atmosphere somewhat and they were able to return to the cabin, throw open the windows and doors, and begin to clean out the dust-filled interior.

A thoroughly wretched Governor was rescued from her cage, which had served to partially protect her from the dust. Her soft coat was combed and brushed.

Everyone was tired and famished, but they had flown without their usual supply of emergency food and water. They were adjusting to the idea of a long hungry night when Blanche caught sight of something in the distance.

"Look everybody, there's a light!" Sure enough, a light twinkled in the darkness, about half a mile away. George shrugged into his heavy army raincoat and prepared to set out alone for what seemed to be a cabin.

A sudden chorus of howling coyotes persuaded Blanche and the girls to accompany him. No way were they going to be left stranded alone out on the lonesome prairie in the dark! So everyone set out, including Governor at the end of her leash.

As they approached the little house they saw it was really a crudely built shack. Four walls, one door, three windows and a roof covered with protruding pieces of tin formed the rickety structure. Abruptly the front door swung open and the rangy silhouette of a man stood outlined against the light. Two women peered nervously over his shoulder. A double-barreled shotgun was pointed right at the Hutchinsons.

"Who's there!" the man shouted, squinting into the darkness.

"Don't worry, it's just a family in distress. I have my wife and daughters with me," George called back. Immediately the gun was set down against the wall as the onlookers focused on the bedraggled foursome before them.

"Well come on in then! What in tarnation are you folks doing out here in the middle of the night?"

The door was pushed open and the family was welcomed into the little house - even Governor. Of course the lion caused her usual commotion, but elicited no signs of fear from these tough people of the plains.

While George and Blanche explained their circumstances, the girls explored the shack. It didn't take long. Everyone lived in the one room. The floor was dirt. A double bed, one cot, a table, and a few odd chairs comprised the furniture. A wood stove sat in the corner. There was no water, or even a sink, to be seen.

Mr. Howlett, their host, was a WWI veteran living there with his wife and her sister. He had been deeded squatter rights to a large acreage of prairie, and was struggling to eke out sustenance from the bare dusty soil.

As was so often the case, those who had little were all too happy to share what they had. Mrs. Howlett fed the hungry brood with bacon and eggs and fresh hot biscuits. The talk stretched far into the night, as Janet and Kathryn grew sleepy and stretched across the double bed. Soon they were fast asleep. Midnight found everyone, except the sister on the cot, stretched width-wise across the big bed, fully clothed and sleeping soundly. Governor, leashed to the leg of the bed, lay as close as she could possibly get to the dying embers of the wood stove. Her eyes, wide and glowing in the dimness, did not close. She was mesmerized by the sight of some thirty chickens roosting all around the top of the room on narrow wooden perches. The scrawny hens clucked nervously and rustled their rusty feathers. A few times the little lion simply couldn't stand it and lunged for one, only to be yanked back to her senses by the leash.

Outside the coyotes howled and the rain continued to beat down on the tin roof. But the Flying Family slept on, exhausted.

* * *

The peaceful interlude was brief enough. Morning brought a rude surprise. A loud pounding on the flimsy door awakened everyone.

"You folks come on out of there!" cried a rough male voice. "Open up and come on out!"

Peering anxiously through the little window, Janet Lee and Kathryn saw a strange sight A dozen or so men armed with long barreled shotguns surrounded the cabin. One man held tightly to a pair of big bloodhounds on a leash.

Mr. Howlett snatched up his own gun and approached the door.

"Who are you people? What do you want?" he shouted.

"We're the law! You're surrounded, so come on out with your hands up."

"It's just like a real western movie," Janet Lee whispered to Kathryn.

George peered through the curtains. "It *is* the law. We'd better cooperate."

"Are they going to arrest us and throw us in jail, Daddy?" Kathryn asked anxiously.

"It's just some kind of mistake. Don't worry," their father reassured them.

"I said come out and come out now!" the voice cried again.

The Hutchinsons and their hosts threw open the door and, hands held high above their heads, filed out into the yard.

The would-be posse was pretty surprised to catch the Flying Family instead of the bank robbers they were tracking. The grounded Stinson had been spied from the air by a search plane, and assumed to be the getaway vehicle. Soon all was cleared up, and the Howletts, the Sheriff, and all his men circled around to wave their hats and cheer as the Hutchinsons took off once again. Onward toward Santa Fe they flew at last, where hot baths, a good meal, and the signature of the Governor of New Mexico awaited them.

After servicing the motor which sorely needed it following the dust storm, they took off for Oklahoma City. East of Amarillo Texas rain clouds formed once again and turned into a heavy downpour.

Flying primarily by dead reckoning meant visual contact with the ground was of utmost importance. Struggling to follow the railroad tracks below at an altitude of barely 200 feet, George knew he would have to depend mainly on his compass and directional gyro. The gas was running low. He needed to hit Oklahoma City dead on.

Rain beat steadily against the windows and steamed up the windows because of the warm air inside. Blanche was busy keeping the windshield and the side windows wiped clear. Finally, the lights of the airport were sighted - not the green landing lights, but red obstructive lights. The plane had to land in the driving rain and sank up to the hub of her wheels in mud.

A tractor and small crowd was gathered to meet the travelers, rain or no. They were amazed that the family had landed in these conditions at an airport that had been closed for three days due to rain and mud. Perhaps divine intervention was suspected. Headlines read "Guardian Angels Bring Flying Family to Oklahoma City" in next day's newspapers.

By midmorning the family had added the signature of Governor "Alfalfa Bill" Murray to the scroll and had another muddy takeoff, aiming for Austin and the vast state of Texas. The adventurous western leg of their trip was almost over.

8

HOMEWARD BOUND

Now the days were rapidly ripening into Spring, and the girls could actually watch, from the air, the mantle of green spreading over the landscape below. Their course turned back eastward as controversial Governor Huey P. Long welcomed them to Mississippi. Arkansas, Tennessee, Kentucky, Ohio and West Virginia followed in short order.

Janet and Kathryn continued to learn firsthand as well as through books about the country they flew over. Blanche taught, but George graded his daughters on their ability to name each state's capital, area, population, weather and climatic conditions, Governor, principal industries and the historical background of the state including battle dates and prominent early settlers. Geographically they were required to know names of principal cities, rivers, mountains and lakes, as well as how high the mountains were, the source and mouth of each river, and the minerals found in each state. As if that weren't enough, they learned to name the stars and identify cloud formations. The things the sisters learned in their Classroom in the Skies were not abstract facts, but dynamically related to the world around them. The gamelike quality of their lessons made them fun. George Hutchinson was nothing if not competetive and fostered that quality in his children. This joy of learning so instilled never left Janet of Kathryn. They learned to strive for excellence and compete not only with others but with themselves.

Blessed with good weather, the Flying Family was covering distances fast and making up for past delays. In one day they were able to cover three southern states: Alabama, Florida, and Georgia. The end was now in sight, and the calm flights between the capitals in the flat Southland seemed easy.

At the Atlanta airport the Hutchinsons were surprised to see an autogyro (an early precursor of the modern helicopter) landing. The strange contraption was still quite a novelty, and the children especially were so interested that its friendly pilot took George, the children, and the lion up for a trip. The four-passenger vehicle was surprisingly comfortable. The children thrilled to the near-vertical takeoffs and landings while George contemplated aloud its future to air travel, intrigued as always by the new and novel.

South and North Carolina came next, with a spring thunderstorm buffeting the plane as they landed in Raleigh. Now only one signature remained to be added, as the family flew toward Richmond, Virginia. Finally this last signature was secured. Everyone was anxious to complete the last lap of their trip and arrive full circle back at Bolling Field in Washington, D.C.

As they circled in for a landing a huge fleet of Army and Navy planes rose in formation and soared around the *Los Angeles*, a huge dirigible that filled the sky over D.C. This air show was unexpected, but provided a dramatic counterpoint to the arrival of the America's Flying Family as they completed their Flight of the States at long last.

The odyssey, with all its trials, adventures, near escapes and triumphs, was over. They had flown over 25,000 miles without injury. They had accomplished their mission and made a kind of history. They had also forged an identity for themselves as a uniquely American phenomenon: a flying family with the kind of popular notoriety and fame that was normally accorded to film or radio stars.

George Hutchinson, some five years after the trip, wrote *Flying the States*, a dramatization written for school-aged children. For a time the book was widely used as a reading text in public schools.

* * *

Looking back at the journey raises several questions as to the judgment calls made by this supremely confident aviator and self-promoter, George Hutchinson. He had flown his family thousands of miles over every conceivable terrain and in all weather conditions. Sometimes his choices added to the risk factors involved.

Why, as asked before, did he take off due north in the harshest winter months, instead of heading south first? Why did he so often choose to fly in the face of storm or snow or foul weather warnings? Why choose to fly in the dark, or without adequate food or water aboard? Why push so hard to make up lost time that exhaustion became a factor?

A picture emerges of a man who loved walking the razor's edge. When he chose the difficult path and blithely ignored the warnings of more cautious folk, he could exult in his ability to pull off another good one. While there is no doubt whatsoever that he loved his family dearly, he also loved playing the hero by coming to the rescue. He had great faith in his own abilities, and it can't be denied that they were exceptional. He also came in for perhaps more than his fair share of phenomenal good luck.

Even in these early years there were many who judged him harshly for putting two small daughters in possible jeopardy. But he didn't see it that way. And these critical voices were drowned out for the most part by the adulation of ordinary Americans who saw in the Flying Hutchinsons life lived large and fearlessly.

Whatever one's opinion of George Hutchinson, it must be said that had he been a more cautious man, the adventures of the Flying Family would never have taken place. And it was, indeed, a glorious adventure.

* * *

Soon it was back to the Governor Clinton Hotel in New York for the travel-weary Hutchinsons. George would spend the next few months speaking to groups and trying to drum up support for future expeditions. Sometimes, the girls would accompany him and tell about their adventures. They were great crowd-pleasers.

In March 1932, the kidnapping and subsequent murder of the baby son of Charles and Anne Morrow Lindbergh horrified and mesmerized the nation. The tragedy struck close to home for the Hutchinsons, shocked into a new awareness of the vulnerability of children in the public eye. The girls were not let out of the sight of adults.

The Canters continued to be great friends and Mr. Canter continued to help the Colonel with introductions to important contacts. Life, for awhile, settled into a comfortable routine at the hotel.

Still, the presence of an African lion in a respectable mid-Manhattan hotel meant attracting, on occasion, more notoriety than the Hutchinsons were aiming for. Enchanted at the sight of the "wild" beast sitting docilely at the feet of a tiny girl in a big chair in the lobby, guests and visitors swarmed to the site.

"Oh, isn't that cute!" a large woman in a fur-collared coat gushed. "What's your kitty's name, honey?"

"What do you feed it?"

"How do you keep it from biting people?"

Questions, questions! Through it all both the Hutchinsons and the lion remained patient and courteous. One day, however, their luck changed

An overbearing woman dripping with furs, diamonds and an air of entitlement sidled up to Governor in the lobby.

"Don't get too close, Ma'am. He's mostly used to just the family" a worried Janet Lee tried to warn her off.

"Nonsense! If he can be trusted with a child I'm sure he's perfectly tame. I just want to give him a little pat." Fingers bejeweled with rings reached out to pat the furry head.

"Ow!" The woman jumped back with a shriek, clutching her forearm. It bled lightly from the raking of four razor-sharp claws.

She sputtered with anger and fear. "What kind of hotel would allow such a thing! You're going to hear from my attorney on this!"

Fortunately, Mr. Cantor intervened, and the matter was settled out of court.

* * *

More controversy ensued when Janet and her father began to play out a daily ritual in the hotel lobby. A decorative carved balcony overhung the main floor. When George came home from his fundraising trips it became a habit for Janet to fling herself off the balcony and into his outstretched arms far below. Janet had absolute confidence that her father would be infallible in this, as he seemed to her to be in all things. As she hurtled through space toward the arms open and waiting for her, the little girl felt no fear. Her father would always catch her. And, of course, he did.

What Blanche may have felt as she watched can only be imagined. Chills of alarm arose in the onlookers in the lobby at the sight of this tiny creature casting herself off a high ledge and catapulting toward the hard marble floors below with all the abandon of a circus trapeze artist. Protests were made to the hotel management.

George Hutchinson had a difficult time granting credibility to these issues. He was not a man who second-guessed himself. He felt his child's trust was absolutely justified and couldn't understand the concern of more timid souls.

At first, the family was actually requested to leave by the hotel. The management didn't want to be held liable for accidents, nor did they want adverse publicity. Yet again, it was Mr. Cantor who intervened and things calmed down after George reluctantly agreed to forgo their game.

* * *

If only Governor had stayed a cub forever. But she grew. And the sight of an adolescent lion strolling down Fifth Avenue at the Hutchinsons' side, even though always leashed, began to excite a new kind of alarm in passersby. Sidewalk pedestrians froze in shock, not quite able to believe their eyes. Some scurried to the opposite side of the street. But whatever the reaction, Governor never went unnoticed.

The mascot, of course, was housebroken. When she needed to be taken outside, she growled softly. Finding places in the concrete jungle to curb a lion, though, was sometimes difficult.

On one of their walks, George and Governor passed an Episcopal church on Fifth Avenue that had wrought-iron palings protecting a grassy area inside. In the process of lifting Governor over the fence to do her duty, the pair were "caught in the act" by an indignant parishioner. The church threatened to prosecute. George got off by paying a a fine instead.

Amid all the excitement, however, life assumed a kind of normalcy on the ground it hadn't had in the air. Janet and Kathryn continued to be schooled by Blanche and also by a governess, Miss Newmeister. Miss Newmeister took the children for "play dates" with other children and their nannies, The girls had had very few opportunities to play with other children in their brief lives, and they relished the chances that did come. The governess also took the children to the New York Public Library, which they adored. Janet felt a special affinity for the big stone lions.

There was a livery stable down the street from the Governor Clinton, and the girls would run down to it whenever they had the chance. The horses were a great attraction, but they also enjoyed playing games with the children of the stable manager.

The months passed, and Governor grew apace. The older Hutchinsons knew that the time was fast approaching when their cherished pet become a real problem. While lion cubs made wonderful pets through their early months, once sex hormones began kicking in with adolescence they could become unpredictable and even dangerous. Recognizing that Governor's time with them would be coming to an end, and a safer zoo home would have to be found for her, still they put off that dreaded day. She was truly part of the family. And she and Janet were inseperable.

As it happened, that decision was made for them. One day, a knock came on the door to the Hutchinson suite. A city policeman, big and burly, filled the doorway.

George knew the jig was up.

"I'm sorry, sir, but the city can't allow full-grown lions to be roaming the streets. I'm afraid I'm going to have to take it with me." He gestured toward Governor, placidly blinking up at the stranger from the carpet in the living room where she had been tussling with Janet Lee.

Janet peeked around her father's legs and saw that the policeman had a heavy chain and a leather leash dangling from his right hand.

" There's a home waiting for your lion in the Central Park Zoo," he added.

George could only try to stall for time. "We've been looking into the matter, Officer. We understand we need to find him an appropriate home."

Behind him, two small faces peered up at the officer in dismay.

"No, Daddy, they can't take him away, can they? Not now!" Janet's voice shook, already full of tears as she pulled at her father's hand.

But George, uncharacteristically, didn't know what to say to his heartbroken youngest.

Desperately, Janet Lee turned to face the tall policeman in blue.

"Mister, you can't take him. He's ours, and he's used to us. You can't take him away and put him in a cage!"

"He's been so good," Kathryn added her protestations. "He won't know anybody in the zoo. He'll be so lonely!"

But the officer had a warrant, and the law was the law. Resolutely he turned away from the stricken little girls to their parents.

"I'm really sorry. I've got kids of my own, but the law's the law. I've gotta take him with me now."

In a daze Janet watched the big policeman fix a chain around her best friend's neck. A last hug, and Governor was led away docilely at the end of the long arm of the law.

Janet's world was changed forever. Her constant daily companion for the past year of her young life was gone. Other pets would come and go in her animal-filled life, but never again would she have a relationship so intimate. The communion between the little girl and the African lion cub had been instinctual and deep.

She was nearly hysterical in her initial, raw grief. Kathryn and Blanche were also weeping. The experience of losing Governor was profound for everyone. A real member of the family had been lost. No more would Janet wake to the rasp of a warm tongue licking her head and ears to wake her up. No more sneaking Gov tidbits of food or hiding her beneath the bedcovers.

The wound never completely healed. Some seventy years later, tears filled the blue eyes as she spoke of the place Governor the Flying Lion had occupied in her heart.

"That lion was the love of my life."

9

TO CROSS THE ATLANTIC

Almost surely, the loss of Governor the Flying Lion would have been even more difficult to cope with had not the excitement of another trip loomed.

George hadn't been idle during their months in the New York hotel. He was anxious to follow up America's Flying Family triumph with the Flight of the States with an even bolder feat. Hutchinson wanted to go Lindbergh one better in staging an attention-getting trans-Atlantic flight. Not only would he cross the Atlantic but would do so with Blanche and the children aboard - thus becoming the first family to cross the ocean. He felt that this would demonstrate that transcontinental flight was not only for aeronautic daredevils but for ordinary people and families.

Lindbergh had flown straight across the ocean. The Flying Family would fly from New York to England by way of a northern route: landing in Greenland and Iceland, then on to the British Isles. There would be stops for refueling along the route. Interestingly, the controversial route proposed by Hutchinson later became a favored course for major airline carriers, proving to be the most fuel-efficient way of reaching northern Europe.

When news of the trip became public, media controversy erupted yet again about the advisability of taking such a risky voyage, north to a wilderness of snow and ice, with children aboard. George Hutchinson sought to divert such critics. In his Foreword to *The Flying Family in Greenland,* the children's book he wrote chronicling the trip, he emphasized that there was nothing more natural than for children to follow their parents, whether for a

Sunday ride in the car or to exotic places. He painted an idealized portrait of a family who shared everything, especially fun and excitement. What he didn't do was directly address the question of danger and risk.

* * *

By 1932, the Atlantic had been successfully crossed several times by other pilots since Lindberg's historic flight, as well as by one woman, Amelia Earhart. Business executive and *Chicago Tribune* publisher Robert McCormack had failed in his expedition to establish an airmail and passenger route from Chicago to Berlin over the Arctic wasteland in 1929. His Sikorsky S-38 Air Boat or "Flying Yacht", the same type of craft the Hutchinsons would use for their Atlantic flight, sank due to ice and high winds in northern Labrador.

Undeterred by adverse press reaction to his new family enterprise, George Hutchinson aquired the backing and the plane for his ambitious trip. *The City Of Richmond*, their new Sikorsky Amphibian, was truly more like a flying boat than an airplane. They would need such versatility for ocean landings and takeoffs. It would comfortably hold the expedition's crew of eight. The four Hutchinsons would be joined by a navigator, radioman, mechanic, and a cameraman. *National Georgraphic,* one of the major backers of the expedition, supplied reel upon reel of film which they expected to receive back full of dramatic footage of the undocumented Arctic wastes for which the Hutchinsons were bound. Little did George Hutchinson realize that the camera film itself would soon become a tool for survival in an adventure that rocketed the Flying Family to international fame and had the world holding its breath to learn of their fate.

* * *

"We're flying to Greenland!" The excited little girls made the rounds at the hotel, telling everyone they met about the trip. Soon the Hutchinsons were deluged once more with reporters, photographers, newsreel men, and others keen for news.

Radio commentators in those days, were highly influential. Lowell Thomas and others criticized Hutchinson not only for the prospect of putting his wife and children in jeopardy by the Atlantic crossing, but for the very route chosen. Why take the new and controversial Arctic flight path, over the frozen wastes of Greenland?

Perhaps the controversy itself fueled public interest in the Flying Family and their new enterprise. In any event, George was undeterred in his optimism and determination. Careful plans were laid, and events began to unfold rapidly.

After taking delivery of the new plane in Bridgewater, Connecticut, Hutchinson proceeded to the Floyd Bennett Airport on Long Island, where it would be hangared until the trip. The new Sikorsky was indeed a marvelous craft. The cabin seated ten people, and was comfortably outfitted. A long boxlike couch covered with a red velvet cushion provided a great spot for the children's naps and had ample storage underneath. A total of six red leather reclining chairs provided additional seating. A soft red rug covered the aisle and two folding tables provided space for eating, study, games, and writing. A lavatory and storage compartments in the back completed the outfitting of the roomy hull.

The Sikorsky's pilot compartment boasted dual controls and an instrument panel with the latest array of navigational instruments, providing an enhanced sense of security. Twin 425 horsepower engines powered the plane. Like a boat's hull, extra storage extended inside the bow. Pontoons were attached to each lower wing to balance the bulk of the heavy plane and make it easy to handle on the water. The main wing measured a full sixty feet across; four gasoline tanks, each with a 330 gallon capacity, filled the wing itself.

The hull and lower wings were painted silver. The upper wing was yellow. George named his spectacular new airplane *The City of Richmond* after the Virginia capitol he grew up in. A decorative crest proclaiming *Familia Volans* (literally, "Flying Family") was painted below the plane's name on the bow of the hull. The crest bore painted silhouettes of each family member circling a three-bladed propeller in the center. The Amphibian under sail was truly an impressive sight.

The carefully selected crew was made up of men who were experts in their fields and adventurous enough to want to be part of an ambitious undertaking. George had handpicked them during the past eight months' preparations. Peter Redpath or Pete, a Canadian, was the navigator. Joseph Ruff, "Joe", the master mechanic, had a thorough working knowledge of the mysteries of the amphibian aircraft. Jerry Altfish served as radio operator and sound technician, with the all-important task of keeping them in touch with the outside world. Norman Alley was a motion picture director and cameraman, on board to create a documentary film of the trip as it unfolded.

Preparations were extensive, time-consuming, and filled with endless details. For the documentation alone, they had to store 15,000 feet of film, packed in cans, as well as a professional movie camera with tripod, several smaller hand-held cameras, and sound equipment along with cables, batteries, and microphones.

Supplies for the crew included concentrated foods and canned items, blankets, first-aid equipment, tools, spare parts, maps, charts and navigation equipment, guns, life preservers, water, cooking utensils and equipment, games and cards, and of course school books for the children. Included in the inventory were several gift boxes, hidden away to surprise the children with later.

Even more difficult than the monumental task of outfitting and storage was the matter of red tape. An impatient George struggled with negotiations with the State Departments of several countries. Air clearance had to be granted. The study of meteorological surveys, of water and air maps, the intricate plotting of courses - all took time and effort.

A huge supply of gasoline and oil had to be shipped ahead weeks in advance, by boat, to the points where the plane would need to land for refueling.

* * *

August 23, 1932

Finally the day of departure arrived. The Hutchinsons and the crew gathered on the field at Floyd Bennet Field on Long Island sound early in the morning. Fuel tanks were topped off, personal luggage was stowed, and all was in readiness. Despite the early hour there was quite a crowd in attendance. Cameras were rolling, reporters were scribbling, and the photographers were snapping pictures. Family and friends had gathered as well to wish everyone *Bon Voyage,* and more than a few tears were shed.

"Are these for us, Mother?" a breathless Kathryn accepted a beautiful bouquet of roses from a fan.

"Where are we going to put them?" asked Janet Lee, struggling herself under an armful of carnations. She could barely see over the top.

It was a beautiful, clear day for flying. At last the huge plane, bearing its load of over 11,000 pounds, rolled down the runway and lifted off before it had gone a thousand feet. The cameramen below kept rolling as the Sikorsky slowly gained altitude. The Flying Family was off to Greenland at last, amid cheers from the ground and their own shouts of joy and laughter. A holiday spirit prevailed.

Everyone settled down for an easy flight of some five hours to St. Johns, New Brunswick, their first scheduled stop. Janet Lee and Kathryn busied themselves arranging a corner for the two new dolls they had received as presents before they took off. Over Boston, several planes buzzed the Sikorsky in greeting, taking photos of the historic flight from the air.

Heading now into a steady headwind that cut their speed down to 160 MPH, they followed the New England coast, enjoying the clear view of rocky headlands, small villages, and green gemlike islands that looked like storybook pictures from the air. Entering Maine they passed over Old Orchard Beach, point of origin for many successful (and unsuccessful) Atlantic air crossings.

Approaching St. Johns, the plane was surrounded by flocks of sea birds flying dangerously close to the engines. Among pilots it had always been a harbinger of bad luck to hit a bird. There were close shaves aplenty as whole flocks headed directly at the Sikorsky, scattering at the last possible moment.

Leaving the coast of Maine and the United States behind, they passed over the beautiful Bay of Fundy. During a relatively quiet lull, Blanche radioed the story of their flight so far to the Baltimore News and also to London, via the American Radio News Station:

In the air. En route to St. John, N.B., August 23 - Well, we're off!

I can hardly realize it, there have been so many unforseen delays in the last few weeks. And there have been those wellmeaning friends who have stood around with long faces and said to me:

"How can you think of taking two little children on such a dangerous flight?"

I have tried not to pay any attention to them.

Janet Lee is dressing her doll and Kathryn is working a jig-saw puzzle, lying on the lounge. The children were anxious to be off, as we all were. But I didn't think they were any more excited than they have been on any of the other flights we have taken. They have lived in the air almost since they were born...

It's almost time for the children to take their naps now...The whir of the propellers make a perfect lullaby.

Suddenly, the girls heard a faint whimper from somewhere in the cabin. A few minutes later it was heard again, and the children became intrigued.

"What is it, Sis? Where do you think it's coming from?" asked Janet Lee.

"I don't know. Maybe from under the floor of the cabin," a puzzled Kathryn replied.

Janet turned toward her father in the cockpit, hands on her hips.

"Daddy, are you playing ventriloquist again and trying to fool us?" Her father was known for his practical jokes.

George held up both hands in the air to protest his innocence.

"Mother, do you think it's a cat? It sure sounds like one to me," asked Kathryn.

As the plaintive sound continued, the whole crew got involved in trying to solve the mystery. Joe started removing things from the top of the lounge seat. The other men, laughing, joined in the search. Finally the lounge seat lifted up. A tangle of supplies, boxes and blankets filled the space.

"Now what?" asked Janet Lee.

"Shhh! Wait til it cries again, then we can tell where it is," shushed Kathryn.

"I'll just whisper. Here, kitty kitty!" Sure enough, a loud "Meow" sounded from the far right corner of the compartment. A square box labeled *Batteries* was lifted out.

Janet sat protectively with both arms around the box in her lap. Kathryn tore the tape from the top and lifted it off. There sat a little gray Maltese kitten with a big pink bow tied around its neck, a rolled-up message attached. Unperturbed, the little creature leaped out of the box and began to purr, pushing its head against Janet's shoulder. Kathryn removed the paper and read out loud:

To the Hutchinson children:
Here's a mascot for you. Call him "Mike," 'cause he's that kind of a kitten, and it's an Irish heart that's bringing you this kitten with all the good luck and happiness in the world to you on your travels.
Your Fairy Godmother.

Enchanted, the girls watched Mike explore the plane from one end to the other. Their excitement at their new pet was made more acute by memories of Governor and her aching loss. Both parents laughed at the antics of the kitten, but George had a heavy heart. He knew that the wellmeaning gift from the secret benefactress would have to stay behind. The laws in the countries the Flying Family would be visiting didn't permit the entry of animals. How would he break this news to his excited daughters?

They circled the New Brunswick town of St. Johns, spying a large welcoming group gathered on the landing field below. However, the crew had removed the wheels from the plane, planning to use it as a seaplane for the rest of the voyage. The landing was made on the St. Johns River instead, with the townspeople making the short walk from the airport down to the landing ramp.

The town welcomed the Flying Family with gracious hospitality. A banquet at the hotel that evening involved speeches and questions. Finally George made his way to his daughter's room, where the girls were getting ready for bed.

"Daddy, where's Mike?" asked Janet Lee worriedly.

"We left him right here when we went down for dinner," puzzled Kathryn.

The girls started to search around the room for the kitten, calling "Here, kitty kitty! Here Mike!".

Their father sat down in a big easy chair. "Girls, come over here. I have something to tell you."

"But we have to find Mike first, Daddy. He's missing!"

"That's what I want to talk to you about, Honey."

Cautiously, still with serious faces, the girls perched on the arms of his chair. They sensed another loss in the air.

"I'm afraid the laws say that we can't take Mike with us into the countries we'll be visiting. He'll have to stay here. Some things you just can't do, no matter how much you want to."

"But what's Mike done? He wouldn't hurt anybody. He isn't even a wild animal like Governor and we flew all over with him," insisted a perplexed Janet Lee.

Their father explained about animals carrying germs, the fear of contagion, and quarantine. Reluctantly they began to accept that their unexpected gift would not be remaining with them.

"What will happen to Mike? They won't put him in jail, will they?" a still-doubtful Janet Lee asked.

"Of course not! In Canada, the laws are much more lenient. One of the customs officers has two little boys, and they are excited to have Mike for a pet, knowing he came to them by airplane all the way from new York. So you need to be happy for Mike, and cheer up," added their father. He was always anxious for happiness, or at least the appearance of it, to be restored. Hutchinsons didn't mope.

Blanche slipped in to tuck the girls in just as they were climbing into bed.

"Mike's gone, Mother!"

She was surprised. "But where has he gone?"

"Daddy gave him away, on account of disease and nicotine," said Janet Lee.

Everyone laughed at Janet's mistaking "nicotine" for "quarantine."

The girls knelt beside the bed and in unison their voices rose in the old familiar prayer of "Now I lay me." Janet Lee added a postscript. "And God, please take care of Mike wherever he is, and let him grow into a big cat with nine lives. Amen."

10

THE ISLAND

The next morning dawned bright and clear, another perfect day for flying. The Hutchinsons and crew were sent off with cheers as they taxied down the St. Johns, gained the air, and lifted off into a blue sky. Anticosti Island, in the Gulf of St. Lawrence some 370 miles north, was their next destination.

Soon clouds gathered, winds increased, and by the time the plane approached the Gulf a hard rain was pounding. George maintained a low altitude. The forests, lakes and islands of Quebec were dramatically visible beneath the roiling black clouds above. Finally Port Menier on Anticosti Island appeared and as the plane touched down within easy reach of the pier, the storm abated. The entire population of the tiny village, some 200 souls, poured down to the bay to greet the travelers.

To their surprise the Hutchinsons found that everyone spoke French instead of English, limiting conversation. Charles Townsend, the representative of the Anticosti Corporation and head man on the island, had sent a horse-drawn carriage to transport everyone to the Chateau Menier. Kathryn and Janet took turns holding the reins as they drove through a thick spruce forest, fragrant after the rain. At the edge of the dirt road the children spied deer and a small herd of elk. The whole big island, they learned, was a wildlife refuge.

Soon the road turned into a long gravel drive that circled in front of a huge, stately residence. Mr. Townsend himself was there to lift Janet Lee down first, then Kathryn, and welcome the whole family to the elegant Chateau Menier, jewel of Anticosti Island.

Opulent furnishings and antiques, shining mahogany paneling, and rich oriental carpets hundreds of years old overwhelmed the travel-weary crew as they entered the mansion. Light filtered through intricate stained glass windows.

The chateau had been built by the wealthy French chocolate manufacturer Menier. A partial tour revealed too many luxuriously appointed chambers to take in. The girls were especially enchanted with an observation room complete with high-powered telescope that provided a birds-eye view of the St. Lawrence river.

Dinner, prepared by a French chef, was delicious. It was two very excited but tired little girls who crawled into their huge four-poster bed that night, heads full of all the things they had seen in their very eventful day.

* * *

Everyone rose early the next day to discover that the island was completely fogged in. Mr. Townsend had arranged for a horseback ride, fog or no, which delighted the girls. On the trail they again spied several of the deer, which seemed to be everywhere. Mr. Townsend told them the herd had proliferated on the little island to a good hundred thousand. Then it was down to the pier to where Pete and Joe were fine-tuning the plane motors while their parents looked on. The girls were joined by two local boys, sons of Port Menier's one and only police officer. The boys only spoke French, but somehow the children managed to communicate with sign language.

Janet Lee spied a motor launch tied to the other side of the pier. Two men were going out to bait lobster traps, and the boys asked if they would take all four children out with them.

"Can we, Daddy? Please?" Janet Lee begged.

"Oh, go ahead," George replied, not giving it much thought.

Soon all six were seated in the launch and the powerful motor carried them out over the still waters. In some fifteen minutes they were back.

"Was it fun, girls?" asked Blanche.

"You bet!" Katherine replied. "And you know something? I bet I could drive that old boat - it looked easy as pie."

"Don't either one of you girls ever try it unless your father's with you," their mother warned.

On the ride back to the Chateau, Kathryn and Janet couldn't stop talking about how easy it would be to drive the launch. As they walked up to the house, Kathryn had her arm around Janet's shoulders and they appeared to be deep in conversation.

"After all that exercise, how about a little rest after lunch?" asked Blanche.

"Sure, Mother." Surprisingly, the girls didn't object to the prospect of a nap, as they usually did.

Lunch was as wonderful as dinner had been: fresh lobster and an exotic French dessert. Sure enough, after luncheon the children went along to their room compliantly. The adults retired to the library to pour over the Chateau's famous collections, enjoying the peace and quiet.

* * *

Upstairs, a very different scene was being played out. "Shhh!" whispered Kathryn, as she led her little sister stealthily out the door of their bedroom. Both girls listened attentively, and heard nothing. Both had their boots, jackets and berets on.

"I think everybody's in the library. Come on!"

Tiptoeing down the carpeted stairs, the two made their cautious way to the back door of the Chateau. Once outside they walked, skipped and ran down the driveway toward the village. "Where are we going, Sis?" asked a breathless Janet Lee.

Kathryn set her lips firmly and squinted straight ahead. "Mother thinks we're still babies, taking naps in the afternoon. Well I know I could drive that old launch, and that's just what I'm going to do."

Janet Lee was silent for a few moments. "But you heard what Mother said! Never unless Daddy was with us."

"I know, but they don't realize I'm old enough to do it. Don't worry, I know what I'm doing."

The launch was still moored at the pier, and the two little French boys came running down to greet their new friends. By pantomime, Kathryn indicated to them that she wanted them to get in the launch with her and Janet Lee.

They shook their heads in alarm, and pantomimed hitting their hands on the seat of their pants - indicating that they'd get a whipping if they got into the boat.

Defiant, Kathryn hopped down onto the floor of the launch.

Janet Lee looked doubtful. "Maybe you'd better get out, Sis. If they whip their own children, they might put us in jail."

"No they won't, because we're visitors. Are you coming, or not?"

Janet Lee hung her head. "No sir! We might get into big trouble."

"Scaredy cat!" Kathryn turned to the confusing array of controls before her. The water was calm, although the fog hadn't lifted much, even at 2 PM. The rope mooring the boat to the pier was still looped around a wooden piling. Kathryn scowled at the instrument panel, and began to have second thoughts.

Without Janet Lee joining her, it looked like she would do best to give up the whole idea. But suddenly, her foot struck the self-starter button on the floorboard. The motor roared, twin propellers spinning, and the launch leaped forward to the end of its tether.

"Janet, help!" Kathryn cried. "Unhook the tow-line!" The boat strained forward but the line held. Janet and the two boys scrambled to tug the line free of the post, and the boat shot away from the shore like an arrow straight from the bow.

As the wind whipped her hair and she headed out into the river, Kathryn was exhilarated.

"I knew I could drive a boat! It's easy," she repeated to herself as she gained control of the steering wheel. She had had some experience in "driving" cars and planes, while she sat in her father's lap.

Cautiously pushing her foot up and down on the accelerator, which she could hardly reach from her perch, she cruised far out into the river. Deciding to head back, she began to make a wide, half-circle turn. When she straightened up, though, she she jammed her foot down too hard on the accelerator and the boat shot forward. To her dismay, she saw that the pier, shoreline, and everything but the water and the fog had disappeared. She was fogged in. Worse, the throttle had become jammed close to the floorboard because of the abrupt way she had struck it. Wherever she was heading, she was racing toward it at full speed.

Panic gripped her. *What should I do?* She felt paralyzed. Thinking fast, she began to steer the boat into a series of wide, gentle turns. But still all remained invisibly shrouded in fog.

Suddenly, a heavier batch of fog dropped down right on the water and seemed to swallow her up, boat and all. In despair, she put her hands over her eyes and let the boat run wild. She was sobbing with her head resting on her arms over the steering wheel when the boat shot out suddenly into the sun.

Meanwhile Janet and the boys heard the motor coming closer and closer, when suddenly the launch hove out of the mist in front of them, headed straight for the pier.

"Boy, Sis sure can drive all right!" Janet Lee cried excitedly. But in the next few moments, as the launch barreled on toward the pier with no signs of slowing, she jumped to her feet.

"Look out, Kathryn! You're going to hit the pier!"

Five hundred feet, four hundred, three hundred, two: the children were jumping up and down with excitement. The boys screamed in French as the launch loomed nearer and nearer.

Just when it seemed inevitable that the boat would crash into the pier, the motor sputtered and spit. A terrified Kathryn raised her head. One word on the dashboard caught her eye: *Switch,* and below it, in smaller letters, *Ignition.* In a flash she remembered her father cutting the ignition switch on the plane, and saying "Always cut the ignition when the motor dies. It's safer." Shouldn't the same thing apply to boats?

Quickly, she turned the switch. The motor sputtered and died. A tremor of relief poured through her body as, some 25 feet from the pier, the boat slowed at last. Turning the wheel sharply to the left, she missed the end of the pier and drifted along toward the mooring she started out from. The other children ran to grab and tether the rope.

Janet Lee grinned at her sister. "You were a great driver, Sis! But you really scared us there for a minute. We were afraid you'd run into the end of the pier."

Kathryn, stepping stiffly out of the launch, prayed her knees wouldn't give way as she climbed onto the dock. *At least nobody knows what really happened.* She tossed her wild, windblown auburn hair out of her eyes and pulled her beret firmly down over her head.

"There's not that much to it, really. You just have to keep your head."

As Kathryn drew close to the others, Janet Lee peered suspiciously into her face.

"You've been crying!"

"I have not! The wind makes your eyes water, is all."

* * *

The girls slipped back into bed with ten minutes to spare.

"Wake up, sleepyheads! You've been sleeping two whole hours!" Blanche jiggled Janet's big toe under the coverlet.

Kathryn yawned and turned her back. "But Mother, we're really tired," she said.

The girls dragged themselves out of bed and rejoined their parents downstairs.

Their host, Mr. Townsend, was away on business, so the Hutchinsons had the run of the mansion. They settled comfortably into the drawing room.

"This is our throne room, Kathryn - the throne room of our castle!" crowed Janet Lee, perching on an ornate chair formerly used by a French king that sat on a pedestal in the middle of the room.

"Shhh, girls, I want to hear this news report," said George, leaning in closer to the big radio set on the table nearby. He adjusted the volume.

The news wasn't good. The announcer reported that three planes had taken off from Floyd Bennett Field on August 23. The first, headed for Norway, was unreported and it was feared that the pilots were lost at sea. Another flight had crashed into the sea off Newfoundland during a heavy fog. A third, America's Flying Family, were temporarily fogged in at the luxurious Chateau Menier on Anticosti Island. This party of eight was the largest ever to attempt a crossing of the Atlantic by air.

"If the Hutchinson family are listening in, we all want to wish them 'happy landings.'"

It helped to hear the good wishes from home, but the misfortune that had overtaken the other flights saddened everyone. A subdued Hutchinson clan said their goodnights early.

If George had worries or misgivings about the perilous unknown into which he was taking his family, he kept them to himself. If Blanche lay awake at night worrying, she had learned to suppress that fear and put a good face on things. The children, bred to view dangerous situations as merely another interesting adventure, and with implicit faith in their father, slept deeply and without fearful dreams.

The next morning George got up early and went to his daughters' room to wake them. Strangely, the girls were still quiet and appeared nervous. George sat down on the bed and asked if anything was wrong.

Janet was the first to break.

"Kathryn, don't you think we'd better tell Daddy about yesterday?" she asked.

Kathryn's head was still burrowed under the pillow. Reluctantly, she sat up in bed, her round cheeks tear-streaked.

Haltingly, the story of the boat fiasco came out. Kathryn took all the blame. "It was all my fault. Janet Lee only went because I made her."

George was taken aback. "You're telling me you drove that launch all by yourself?"

"Yes, Daddy."

George sat on the edge of the bed and put his arm around his daughter. He explained the danger she had put herself and others in, and also the wrongness of taking someone else's property without asking. Tears were dried, and it was a brand-new morning.

George's descriptions of the mischief his daughters got into, however, had about them an unmistakable ring of pride. His daughters might have been disobedient at times, but clearly he admired the resourcefulness with which they usually got themselves into, and out of, trouble. They were chips off the old block.

* * *

The weather went from bad to worse over the next day, and the Hutchinson party found themselves still unable to take off. The group decided to get busy that afternoon, rain or no rain. The four men went lobstering, and the Hutchinsons donned rain gear and tried their hand at trout fishing in a nearby stream.

The canny trout seemed bound and determined to elude them all, except for Janet Lee, who caught fifteen in all. It was trout for dinner that night. It was trout for breakfast and lunch the next day too, which at least brought clearing weather.

The adults busied themselves filming their beautiful surroundings and life on the island. The elk herds were an attractive subject, but became frightened when the humans and cameras ventured too near.

Janet Lee the animal lover protested, "Please don't scare them, especially not the women elks!"

The girls wandered and explored the grounds around the Chateau. Janet Lee spied a little fawn standing in the dappled sunshine at the forest's edge.

"Look, Kathryn! Don't breathe! Isn't he cute?"

"Maybe we can catch him," replied Kathryn, starting to walk slowly toward the fawn.

But the fawn turned and hobbled toward the woods, limping badly as though one of his legs were injured.

"He's crippled! We have to help him," cried Janet Lee, and took off running after him.

Injured or not, the baby deer proved too quick for them and disappeared back into the forest. Hesitant, the girls paused and peered into the gloom ahead of them.

"Where is he?" Janet whispered.

"There, right there, behind that big rock! You go left and I'll go around to the right, okay?"

Creeping softly, the girls circled the rock and sure enough, there was the little creature crouched down on the ground. As his would-be captors lunged at him from both sides, the fawn sprang forward and darted deeper into the woods. Janet and Kathryn sprawled empty-handed on the soft needles of the forest floor.

Quickly they scrambled to their feet and took out after their elusive prey once more. Every time it looked certain that they would be able to overtake the fawn he always bounded away, just ahead of them. It was tantalizing. Once they even touched his soft fur.

Soon the girls had lost awareness of anything except their single-minded determination to catch that crippled little creature. Deeper and deeper the cathedral-like woods opened before them, the trees so tall and dense almost all the sunlight was blocked out. The dappled play of light on the floor of the forest sometimes fooled them into thinking they were glimpsing the fawn's spotted coat. Fifteen, twenty, thirty minutes passed.

Finally, panting with exertion as they leaned against a big rock, the girls decided to call it a day. Suddenly they noticed how dim it had grown around them during their chase. They could barely see the sky above the looming dark shapes of the tall trees. Discouraged and tired, they began doggedly to retrace their steps back to the Chateau.

It didn't take long for them to realize that everything looked alike, and they couldn't remember what direction they had come from.

"What'll we do, Sis?" Janet whispered.

Kathryn attempted to at least look brave. "We'd better just keep walking in one direction. After all, this is an island. We're bound to reach the water sometime." She did not share with her little sister the fact that according to their host, the island was some 300 miles square.

The tired girls walked and walked and walked some more. Occasionally, they altered their direction. The sun slanted through the trees at a lower angle. Five o'clock passed, then six.

"I have an idea," Kathryn said. She knew they had to do something before it got dark. "We'll climb to the top of one of these trees, and maybe we'll be able to see the Chateau and where we are."

Kathryn started to climb the tallest tree she could find. The climbing itself was easy as there were so many branches for handholds, but the sticky sap stuck to her hands and clothes and shoes. At the swaying top, she was unable to see over the other trees. They were adrift in an endless sea of green.

A subdued Kathryn climbed back down. Now both the children were beginning to be apprehensive. Every little sound was magnified as they walked along, holding hands for comfort.

Every few minutes they shouted as loud as they could:

"Help! Mother, Daddy! Help!"

But the shadows kept growing deeper, and at last Janet Lee started to cry.

"It's getting dark! And what about all the wild animals?

Kathryn sounded braver than she felt. "Don't be silly, there aren't any animals like that around here. Just deer, and elk, foxes, maybe wolves."

Both girls grew silent at the thought of wolves.

Then Kathryn brightened as another idea came to her. "I know how we can stay safe even if we have to spend the whole night in this old forest."

"How?" Janet Lee's voice quavered.

"We'll climb a tree and lash ourselves to it with our belts. That way if we fall asleep we won't fall off."

So as complete darkness enveloped them the girls found a perch on the branches of a tall spruce, and lashed themselves to it with their backs against the trunk. Night in the forest grew deep all around them.

* * *

Meanwhile, Blanche had returned to the Chateau after the filmmaking to find both children missing. No one had seen them for hours. Worried, she sent the servants out from the Chateau in all directions, searching and calling.

When George and the others came up the driveway with the camera equipment, she ran to her husband.

"George, we can't find the girls anywhere! They're gone!"

Within a short span of time, all the available men from the township and the Chateau were out in force with lights and guns, spreading out to search the woods. Mr. Townsend showed up with an additional twenty men.

An idea flashed through George's mind. Working at a feverish pace, he readied the plane for a water takeoff, gave her full throttle, and lifted up over the woods.

The men below had instructions. When they saw a bright light drop from the skies, they were to make their way toward it, build a big fire wherever it fell, and keep it burning.

Meanwhile Kathryn and Janet had heard the gunshots in the distance. They renewed their shouts and calls, to no avail.

Suddenly they heard the reassuring roar of the Sikorsky's engines just over their heads and glimpsed the bright landing lights.

"It's Daddy! It's going to be all right!" Instantly Janet Lee's apprehension turned to joy.

"Look, he's dropped a light over there! It's a flare! Come on," said Kathryn.

The girls unhooked their belts and scrambled back down the tree. They made their way as fast as possible toward the burning flare, which had landed a few hundred yards away.

Running through the darkness, Janet Lee could see the bright flare just ahead, when suddenly she tripped and fell. Something moved beneath her. She screamed.

Kathryn helped her sister up, and the flare illuminated the form of the little fawn, the same one who had been the cause of all the trouble, lying curled in the leaves and trembling with fright.

From the air, George and Mr. Townsend watched the search parties with their lights make their way toward the burning flare on the ground. Suddenly the hand-held flashlights began to circle wildly as the rescuers signaled that the children had been found.

It seemed to take an eternity for the anxious father to return to the pier, land and moor the plane and make his way back to the Chateau. Just as the automobile reached the turn in the road, itsr headlights illuminated the figures of Janet Lee and Kathryn emerging from the woods carrying something. The entire search party trailed behind them.

The girls broke into a run when they spied their father. Dirt-streaked faces shone with joy as they lifted up their prize.

"Look, Daddy! We've got a baby deer! A live baby deer!"

II

LABRADOR

August 30th brought somewhat unsettled weather, but all were determined to press on northward with the flight. Almost everyone in Port Menier lined the pier to see the expedition off. Farewells were said to new friends, and Mr. Townsend promised to send the girls news of the fawn by radio, He was nursing it back to health following its stressful night in the woods with the two sisters.

After a routine takeoff, it was business as usual with Pete watching the maps and setting the course, Jerry manning the radio and George of course at the controls. The children and Blanche occupied themselves with trying to fill in an outline map of northern Canada with the lakes, mountains and rivers they were flying over. Gradually, as they flew over the wooded, lake-dotted interior of Quebec and crossed into Labrador, the forest began to thin out. Barren wasteland and rock stretched out endlessly below them.

The North Atlantic with its numerous inlets and rock islands hove into sight, but it was hard to make the irregular coastline jibe with their maps. They were excited to spy their first icebergs, still far out in the ocean.

"Watch out carefully," said George. "We don't want to miss Hopedale." The small village appeared as a tiny dot on the complicated map. Jerry, in radio contact with the Newfoundland radio station, reported that the weather in Hopedale was perfect, with an unheard-of high of eighty degrees.

Finally spotting the tiny village, the plane began to descend. They were now north of the tree line, and only a few scrubby bushes relieved the severity of moss-covered rock in all directions.

Downtown Hopedale consisted of the general store, a building that originally belonged to the Hudson's Bay Company, a church, mission house, and radio station with high tower. The rest of the village was cluttered with dilapidated wooden huts. Husky dogs lay sprawled between the little houses.

As the Sikorsky circled the deserted-appearing village, it sprang suddenly to life. Inuit of all ages and sizes came rushing down to the water's edge as the plane approached. Dogs barked wildly and ran around in circles. After a smooth landing, the plane taxied to the pier and everyone went ashore.

Greetings from the native population were warm and enthusiastic. They pumped the hands of their visitors vigorously, wide smiles spread across their faces. Mr. Perritt, a Moravian missionary, with his wife and daughters officially welcomed the group. Along with the storekeeper and radio operator, they made up the entire Caucasian population of the town.

The girls stared in fascination at the Inuit, many dressed in sealskin boots, pants and shirts. The native children stared back, equally intrigued by these strange new visitors who had swooped down from the skies and especially by Janet Lee and Kathryn.

The eight travelers were welcomed by the Perritts to the Mission House and made comfortable. That night they enjoyed a nutritious if bland dinner based almost exclusively on canned food. It was a far cry from the French cuisine of the Chateau Mernier. At dinner they learned about the missionaries' struggle to provide food, clothing, and health care for the Inuit population in this extremely poor region.

It was obvious to the Hutchinsons that in the white man's attempt to "civilize" the native peoples, they had created dependency instead. Here the old ways of the hunting and fishing traditions struggled to coexist with modern life.

A tour through the town revealed as many as twelve people sharing a single shack. Everyone from grandparents to grandchildren slept in one room on raised planking piled with furs for bedding. Holes in the ground furnished sleeping quarters for the dogs, where they curled up together in one big ball for warmth. These dogs' lives were nothing like that of pampered American pets.

They were bred for hard work, and lived on refuse thrown away from the natives' homes.

Mr. Stevenson, operator of the radio station owned by the Newfoundland Government, showed them around the station. He had already radioed the press of the expedition's safe arrival. What a surprise it was, in that back-of-beyond place, to hear the voice of popular announcer Lowell Thomas telling the world about their progress.

As they prepared for bed a weary family gazed up at skies that seemed to be filled with a million stars. The milky way looked close enough to touch, and a brilliant half-moon illuminated a landscape as alien as the moon itself. The clarity and beauty were overwhelming. George noted that the thermometer had dropped from a high of 80 earlier that day to 28 degrees Farenheit. All slept soundly that night in spite of the howling of the huskies all around them.

* * *

The next day was set aside for motion picture work. Documenting life in the Inuit village and getting shots of the Sikorsky landing and taking off were priorities. The entire population, fascinated with the camera equipment, gathered round eager to watch, help, and play a part when they could.

In the meanwhile Janet and Kathryn, not needed in these scenes, wandered down toward the pier. Today was cold enough for them to wear their fur coats warmly buttoned up over their riding habits. A rim of ice had formed along the edges of the shore during the night, and huge icebergs drifted in the inlet. Casting around for something to do on the pier, the girls noticed a stand of fishing lines with two hooks on their ends.

"Why do they have two hooks? And where's the bait?" asked Janet.

"Remember what Preacher Perritt told Pete and the others about fishing last night?" Kathryn replied. "There are so many fish in this water, all they have to do is let down a line until it rests on the bottom. Then they pull it up really fast, and the fish bite in a couple of seconds! It's called jigging."

"Well come on then, let's jig!" Janet urged.

Kathryn squinted out over the water. One long iceberg had jammed against the point of land at the edge of the inlet.

"The water's really shallow around here. See, you can see the bottom. The fish are out where it's deep," she calculated. "But if we climbed out on the iceberg over there at the point, we could fish from there."

Janet Lee was thrilled at the idea of fishing from a iceberg, and the two girls lost no time in running along the water's edge until they reached the massive piece of ice that had floated in to shore.

"Watch out, it's slippery!" Kathryn called to her sister, who was scrambling ahead of her toward the far end of the berg that jutted some 200 feet out into the sea.

"I won't fall!" Janet called back. "And even if I did the Eskimos would save us."

Soon they reached the flat part of the iceberg at the end. The larger part of the huge mass now rose behind them, shutting off their view of the shore.

"I'm going to catch the first fish!" boasted Janet, swinging her line to throw it into the icy waters.

"Bet you don't!" Kathryn quickly followed suit. Engrossed, the two girls practiced jigging their lines up and down.

"Ow!" Suddenly Janet Lee's line went taut in her hands. She struggled to hang on as something on the end pulled strongly.

"Pull it in quick!"shouted Kathryn.

Arduously, hand over hand, the little girl wrestled the heavy fish to land, or rather to the ice. It was a 5-pound cod, which flopped about energetically as she heaved it over the edge of the floe. But she froze at the idea of taking the hook out of his mouth as her father had shown them. The still very much alive fish was so big, so wriggly, and so wet.

"Hey, I've got one too. Oh boy, this one is pulling so hard - help me pull!"

On the count of three both girls heaved mightily and not one but two big cod flopped over the edge, caught firmly on the prongs of the double hook.

Together they devised a method to deal with the fish hook removals. Janet Lee stood on each fish in turn, while Kathryn wriggled the hook out of its mouth. No sooner had they turned back to the water for more victims than Janet spied something strange swimming straight towards them.

"Is it a seal?" she wondered aloud, then corrected herself. "No, it's a dog, one of those Eskimo huskies."

The girls coaxed the dog on through the freezing water and helped him struggle over the edge of the iceberg. To show his gratitude he shook himself violently, showering them with icy water. Then he pounced on their hard-won fish, and devoured them in a series of huge gulps.

"Well, there's not much use fishing for more. He'd just eat them too", Kathryn sighed. "Let's go on back and see if we can help with the movie-making."

As the girls scrambled back over the high ridge that had hidden them from sight of shore, they were stunned at the sight that greeted them.

"We've broken off! The iceberg has broken off, and we're drifting out to sea," cried Kathryn.

In point of fact, their floating fortress was already well out into the bay and being carried further from shore each moment by the swift current.

"Help! Help!" both girls cried at the top of their lungs, hoping to attract the attention of the tiny figures they spied moving around back on the beach. They looked so far away. The dog, replete with his fish breakfast, added his barking to the commotion. But it was no use. Tthe wind was against them, and not a single soul on shore heard their cries or turned in their direction.

Suddenly Kathryn brightened. "I know what we can do. The dog! We'll make him swim to shore."

"How will that help?" wailed Janet. "He can't talk!"

"I know. But we'll tie our handkerchiefs and our hats around his neck with our belts. We'll push him in the water, and when he swims to shore and people see those things around his neck, they'll know they're ours, and start looking for us."

When they had lashed their things around the husky's furry neck, coaxed him to the edge of the sea and pushed him in, of course he immediately tried to scramble back on board the iceberg.

"Go home! Go home!" they screamed at him, waving their arms and making throwing motions. Finally the poor dog turned toward shore, a good mile away. He would have to make shore, return, or drown. And still the iceberg continued to drift inexorably toward the open sea.

* * *

Back on land, the crew members were busy staging scenes for their documentary. The Inuit were proving to be good performers, patiently repeating one scene after another for take after take. George and Blanche took a hand camera with telescopic lens down to the airplane, took off, and began shooting gorgeous scenic views out the windows from ten thousand feet. The air was so calm George could fly for up to fifteen minutes with his hands off the controls, snapping pictures from different angles.

As they finally circled back toward the village, they noticed a group of boats way out in the inlet. Three launches were trailed by some twenty kayaks and dozens of people clustered on the pier, gazing seaward. Leveling off to get a shot of this, George thought he saw something move on a big iceberg just below.

Meanwhile the children had anxiously watched the husky's progress toward shore, the growing swells sometimes hiding him from view. Time seemed to inch by. Wouldn't he ever reach land? At long last they watched the dog scramble ashore and disappear over a small hill. When they spied the fleet of boats setting out from the pier, a huge feeling of relief surged through them.

Overhead, the roar of the Sikorsky's motors filled their ears. The plane circled and banked, diving almost straight down upon them.

"It's Daddy!" shouted Janet Lee.

"He sees us! Everybody's coming for us now," cried Kathryn.

The Sikorsky made a smooth sea landing just behind the iceberg and taxied up to the flat part of the ice. Maneuvering a wing close to the ice, George called:

"Come on, you two hooligans! Get off that ice!"

Climbing over the wing and into the hatchway to the cabin, the girls were scooped up in Blanche's waiting arms. Reprimands were lost in the relief of the reunion. Anxious as they had been, the two parents couldn't help but admire the fast thinking that had turned a Husky into a rescuing angel.

The rest of the story was told over dinner later that evening. The dog, after reaching shore, had made his way immediately to the film site. One of the men finally discovered the wet belongings hanging around his neck. Scanning the inlet through a telescope, they spied the children on the iceberg and were setting out to rescue them when the plane touched down.

When the children said their prayers that night, George and Blanche heard Janet Lee ask:

"Dear God, please be extra good to the Eskimo dog who saved us. And if you can find any way to do it, please let us have him for our very own."

Kathryn added, "Amen!"

* * *

On the third day of their stay in Hopedale, the Hutchinson party enjoyed a meal of fresh ham, a true delicacy in this northern habitat. Three hams had flown with them from the Chateau Menier, welcome gifts indeed. This one didn't last long shared by thirteen hungry people. Afterward, the meaty ham bone was given to the girls so that they could present it to the dog who had helped so greatly in their rescue.

The big dog, obligingly enough, was lying just outside the mission gate. He was on his feet and wagging his tail furiously as the scent of the ham bone reached his nostrils. Happily the girls unwrapped the paper, then stood back as he set on it ravenously.

For the rest of the day, the dog, dubbed "Bones" by Janet Lee, trailed behind the girls wherever they went, From time to time he would lay it at the feet of one of the children, back up and bark, and look at them expectantly.

"I believe he's trying to tell us to put some meat back on that bone, Sis," said Janet Lee.

The girls made their way down to the shore once more, where the natives had left several sealskin kayaks bobbing in the shallow waters offshore. All was deserted. Apparently everybody was once again immersed in the filmmaking.

Kathryn contemplated the bobbing boats. "It looks so easy, I know we could do it. All you have to do is kneel in the kayak and work the paddle."

Janet Lee remained skeptical: "But they look so tippy and light. What if we fall out?"

But Kathryn was already kneeling in one of the vessels. "Come on. We'll stay right by the shore."

Doubtfully, Janet followed her big sister's lead and soon both girls were paddling, first awkwardly and then with more assurance, in the shallow water. Bones sat on shore, whining softly. Then he would pick up his bone and run a short way down the beach, drop it, and bark.

"He still wants more meat put on his bone!" laughed Janet Lee. "Whoops!" As soon as she took her attention off her precarious balance, the tipsy boat rocked.

Suddenly there was a splash from shore. Dismayed, the girls watched Bones swimming straight for the kayaks. The ham bone was still clutched firmly in his jaw.

"No! Go away, go home," cried Kathryn, afraid the dog would try to climb in the boats and upset them. He made a beeline straight for Janet's kayak.

"Hit him over the head with the paddle!" Kathryn yelled.

But Janet Lee, of course, would not. Before she could stop him, Bones put both forepaws on the bow and tried to hoist himself up.

The kayak rolled, and with a tremendous splash the little girl hit the water. The shock of the cold was almost paralyzing.

Struggling and sputtering, Janet Lee fought just to stay afloat in her heavy clothing. Although a strong swimmer, she could make no progress toward shore. Her kayak had drifted away. But she threw both arms around the husky's strong neck, and together they plowed through the surf toward land.

In the meantime a badly frightened Kathryn had been frantically calling out for help. Half-dragging her little sister from the freezing water, she threw her own fur coat around her. Janet Lee was silent, except for teeth that chattered like castanets, as the two girls struggled as fast as they could toward the mission. A subdued Bones trailed along behind them. On the path they met their parents, who were emerging from an Eskimo hut carrying a bearskin they had just purchased.

It made a good wrap for their shivering daughter as they scooped Janet Lee up and hustled her back to the mission and directly into a hot bathtub. She was tucked into a bed piled high with covers and given hot brandy and water.

Of course the children chattered all during these proceedings, trying to divert a scolding. It didn't work.

Janet Lee heaved a great sigh. "I guess you're right. Maybe someday I'll learn."

As her father made his way to the door she added, "But please don't be mad at Bones! I love him. And can we give him another ham bone? He dropped the one he had in the water."

12

GREENLAND

The Flying Family's last day on the North American continent dawned overcast and with a promise of rain. Nevertheless, spirits were high as the expedition focused on the next leg of their Atlantic crossing. It was on to Greenland.

Preparations had been extensive for this crucial leg of their trip. Seven hundred miles of open ocean lay between them and Godthaab, the capital of Southern Greenland. Goodbyes were said to the missionary family, who had become dear friends, and to the native Inuit, who had impressed the Hutchinsons greatly. Facing hard lives, with no luxuries or exposure to the supposed advantages of civilization, still the people were loving, honest, content, and peaceful.

Sixty extra gallons of gasoline were stored aboard the Sikorsky along with extra food and water. Everything had been checked and rechecked, overhauled, and reloaded.

The Amphibian, bearing up well under her extra heavy load, sent up waves of spume as she lifted off from the inlet waters. The whole village had gathered to see the travelers off. Soon they were tiny dots far below as the plane rose to an altitude of three thousand feet. Everyone was in a happy, almost carefree mood as they began the most adventurous part of the journey.

"Why don't you fly lower off the water so we can estimate wind drift better by sighting on the drifting icebergs?" Pete suggested.

After adjusting for this, they leveled off at twenty-five hundred feet, only to be souped in by thick fog. Upward they plowed, only emerging from the fog at some five thousand feet. They were now almost a mile high. Here, the sun was bright and the sky a crystal blue.

Joe kept busy pumping gasoline from the extra cans into the main fuel tanks. Pete could only estimate their course as they flew above the fog, setting a general compass course. Norman was setting up cameras for interior shots, and Jerry established communications with Hopewell. Hoped-for signals from Godthaab were nonexistent, so the group did not know of weather conditions there.

Kathryn and Janet Lee took advantage of the lull to climb through the pilot compartment window and sit with their father. Janet sat on George's lap, Kathryn beside him.

"Let me steer, Daddy," Janet Lee reached out eagerly for the big wheel and grasped it rigidly in her little hands.

"But what will *I* do?" Kathryn asked disappointedly.

"You're going to correct your sister by keeping track with the instruments," her father replied. Soon he had both girls totally engrossed in the art of flying. The air was smooth and the big plane responded easily to their efforts.

"Flying is easy," Janet boasted. "Mama, look at us, we're flying the plane by ourselves!"

Blanche poked her head and shoulders through the compartment window.

She grinned at her husband. "Watch out, or those girls will take your job away from you."

* * *

George noticed that the fog bank was piling up with clouds as well. Higher and higher they climbed, up to ten thousand feet, then twelve. Still they were engulfed in a white sea, flying blind. Neither water or land had been visible since a few minutes outside of Hopedale. Outside temperature plummeted to fourteen degrees. Ice formed in the carburetor caused the motors to spit and sputter occasionally. Arctic conditions were asserting themselves with a vengeance.

Finally, Pete received a radio transmission of weather conditions ahead:

Godthaab clear, high scattered clouds. Rising barometer.

George had been afraid to fly under the clouds until receiving the Greenland weather report, cautious of crashing into one of her high mountains as they descended. Now he began to ease the plane into a steep descending glide.

Instantly, both engines started to miss. The instrument dials, perhaps clogged with ice, went haywire. George realized they were falling too rapidly and pulled back sharply on the wheel. Both engines roared to full power once again, causing a sudden power stall. Finally, the air speed jammed.

The plane was falling off on the right wing. Now they were not only plunging downward, but spinning around and around.

"Look out for the water! Watch and tell me when you see anything at all!" George shouted, frantically trying to regain control.

All the crew members were plastered against the windows, trying to glimpse something, anything, through the swirling fog and clouds.

George opened up the throttle full thrust, and the motors sputtered yet again. The wild ride downward would be temporarily arrested when one or both motors leaped back to life. Then they would roar or almost quit. For what seemed like an eternity, chaos reigned. From eleven thousand to eight thousand to sixty-five hundred feet, ever downward they plunged.

Suddenly, shouts rang out from the cabin. "Water, Daddy! There's the water!" The girls had been first to sight the ocean below.

The right side of the plane was pointed skyward in relation to the sea. Immediately George righted her and dove for the rift in the clouds. Fortunately, both motors responded this time. Glimpses of the ocean still alternated with blind white fog as they dove down, down and finally out into the open.

George mopped his dripping face with his handkerchief. Air temperatures became much warmer as they leveled off above the water at about five hundred feet. The gage registered forty-four degrees Farenheit and the instruments, except for the air speed indicator, seemed to be functioning normally once more.

"Boy, Daddy, I was really dizzy there for awhile!" gasped Janet Lee.

"I think we all were," her father replied quietly.

* * *

Soon Blanche was serving sandwiches and drinks, and everyone's spirits rose. Kathryn spied a school of whales, but before Norman could get into position with the camera they were gone. Patches of blue appeared through the clouds and the fog had completely dissipated.

At last there was time to appreciate the incredible vista opening before them. Greenland's mountains rose magestically from the sea. The high, jagged crags covered with ice and snow were bisected by fjords that wound their torturous way down to the ocean. There were no trees visible. Icebergs in fantastic shapes, large and small, dotted the waters around the shoreline. The sight of this huge white land mass illuminated by the sinking sun settled deeply into the hearts and memories of the voyagers. They were breathless with the beauty of it all.

Pilot and crew discovered that in all the commotion they had missed their mark of Godthaab only by a little. Following the coast northward, they soon spied the little village far below. Kathryn and Janet knelt on the lounge seat behind their father and watched the landing in silent excitement.

"Girls, when we land you two will be the first children, and we will be the first family, to fly to Greenland from New York. Just like the pioneers in their covered wagons, we're making history," George said with satisfaction.

Like Hopedale, Godthaab was built at the foot of a mountain range, encircled by a wide bay that provided a safe harbor for ships. Now, in summer, there was no snow except in the mountains. Huge icebergs floated lazily with the tide in and out of the bay. A brilliant emerald green reflected back where their bases disappeared under the blue water.

Taking care to avoid the floating ice, George made a smooth landing and taxied toward the wharf. Danish officials, who happily spoke English, and several native Greenlanders escorted them ashore in their launch.

The crowd of sealskin-clad Inuit on the wharf shook hands enthusiastically and reached out tentatively to touch their guest's strange clothing. Danish families, mostly citizens of Copenhagen assigned to the outpost, greeted them warmly as well.

The Governor of the territory and his wife escorted everyone to his white house on the hillside. They were accompanied by the thong of friendly Greenlanders, who remained outside the house until everyone came back out to shake hands once more with each and every one of them. These poor but hospitable people had brought gifts for their visitors. Dolls, beaded scarfs, fur hats and mittens charmed Janet and Kathryn.

The bare bedroom floors were strewn with dogskin rugs and the family kept warm that night under deep eiderdown covers. The "boys", or flight crew, were quartered on the Danish warship *The White Bear,* anchored nearby.

Friendly welcome aside, the next day held a bitter surprise for the visitors, forcing them to turn in a new direction that would lead them to the very brink of disaster.

Governor Svanne, chief authority over the Danish territory of Greenland, had been notified by the home office in Copenhagen that the Hutchinson expedition had not been granted permission to visit Greenland. Obviously, this was unwelcome news to their host as well as the visitors, but the Governor had no choice in the matter. It was a complicated situation.

George had requested permission to visit all three territories controlled by Denmark: Greenland, Iceland, and the Faroe Islands. Permission was granted for Iceland and the Faroes, but the green light for Greenland had been delayed. Although Hutchinson had been reassured by his New York representative that all was in order, evidently there had been a miscommunication.

Never a patient man, George was perplexed and upset. He knew that other pilots, in inferior aircraft, had been granted permission in the past. Nor had he been given to understand that Greenland was essentially a closed country, open to visitors by permit only.

After a flurry of radio messages between Governor Svanne and the Greenland Commission in Denmark, it was decided to fine the Hutchinsons 1000 kroner ($180). George paid the fine, and expected the matter to be over and done with. It was not.

Now George was informed that the expedition would not be allowed to fly the 463 miles directly across the top of Greenland. They must instead fly 1,200 miles along the coast. He was aghast.

The ice cap had been successfully traversed several times already by smaller planes with less equipment. Flying over the ice cap, ten thousand feet in the air, was infinitely preferable to tracing a treacherous route along the foggy coast. Yet the officials remained implacable to logic.

The reason for the refusal appeared to be the children. In vain did George attempt to explain that his family had practically lived in the air. Nor did the logic that the shorter route would be far less perilous than the longer coastal flight bear any result. With many thousands of air miles to their name, the Flying Family was now being refused permission to fly less than five hundred miles.

George would have been considerably more upset had he known what lay ahead.

* * *

The visit in Greenland itself, however, was filled with the warmth and hospitality of her people. There were no dogs in Godthaab, but there was a blue fox farm and a rabbit farm. Housing and the general health of the natives was superior to that in Hopewell. So was the food. The visitors enjoyed delicious Danish meals at the Governor's table.

An escorted tour of *The White Bear* was fascinating. The children loved playing with the cannon and other weapons on her huge deck, and the sailors enjoyed the presence of the children after being away from home port for some nine months. Captain Evers agreed to transport three hundred gallons of gasoline to Julianehaab for the Hutchinsons, since they would not be able to fly all the way around the coast without refueling.

Over the next few days the girls made friends among the local children and even learned a few new Inuit and Danish words. The lives of the native population had not changed, essentially, in generations. Of course there was no gas or electricity. Janet and Kathryn suffered the indignity of taking cold, standing baths in big barrels, with Inuit girls pouring buckets of water over them.

George remembered the surprise gift boxes, still on the plane, and with the help of two native boys brought them to the house.

"What is it?" the girls asked eagerly, fumbling to open the first box.

"Balloons shaped like animals!" cried Janet, beginning immediately to blow one up.

"Look! An elephant!" "I've got a camel." "Mine's a rooster." The girls were delighted.

Soon the second box was opened as well, revealing piles of candy, big all-day suckers, and still more balloons.

"But how will we ever eat all this candy?" wondered Kathryn.

"They're not for you, greedy. They're for all the Greenlander children," George replied.

"Can we give them to the children? Please?" The girls begged.

"You sure can, but first there's lots of work to do." George's eyes twinkled. "Let's see how full of hot air we all are."

Each member of the family huffed and puffed for the next two hours, blowing up all the animal balloons. When they were finished, some three hundred exotic beasts and birds bobbed and floated about the room.

George and Blanche loaded the girls down with balloons and candy, and sent them out to provide a Christmas in August to the Greenland children. Soon not only the delighted children but whole families swarmed around the Governor's house. The polite native children did not scramble for their gifts, but waited politely to be offered them.

As Janet Lee and Kathryn rushed back in for their last loads, Blanche admonished them:

"Save some back for the children in the next village, girls."

"But Mother, I need more!" Kathryn protested.

"Me too!" added Janet Lee.

This didn't add up. "Why in the world would you need any more? Everyone has plenty!"

"But there's a beautiful scarf the lady wants to trade me!" Kathryn said.

Horrified, George and Blanche ran out into the yard to discover piled up at the gate a huge heap of furs, scarfs, beads, arrows, hats, gloves, and other odds and ends from the homes of the locals. Janet Lee and Kathryn had been trading for the balloons and suckers, not giving them away.

They were proud of their "take". It took a long time for the Hutchinsons to sort out the owners of all the items and return them, but it was eventually accomplished. The Greenlanders themselves remained happy. Nobody wanted the party to end.

Still, when it was all over, Janet Lee sighed wistfully, "It sure was a lot of work for nothing."

* * *

Final authorization from Copenhagen for the resumption of their flight still hadn't arrived when the whole family had to be thoroughly examined at the hospital as part of visitor screening. Bad weather around Julianehaab further delayed departure plans. Each day, the Hutchinsons settled further into the life of the village.

On a fine afternoon the girls got permission to do a bit of exploring. Soon they were scrambling up the steep side of a cliff. Janet Lee, nimble as a little mountain goat herself, reached the goal first: a ledge of rock that jutted out from the mountain. Panting hard, both girls rested and took in the vista before them.

"Don't get too close to the edge of the rock," Kathryn warned.

"Heck, I won't fall. See that big bird's nest?" Janet replied, craning over the ledge and pointing to a crevice just below.

"It's got eggs in it, too," said Kathryn. "I wonder what kind of a bird could have a huge nest like that?"

"I don't know, but I can get one of the eggs. Maybe we could tell."

"You'd better not! What if the mother bird sees you?"

But Janet Lee wouldn't be deterred. "Hold onto my pants leg," she commanded. Stretching over the edge as far as she could, she plucked one of the eggs from the nest. "I've got one! Pull me back!"

Standing on the brink of the crevice, both girls examined the egg.

"You didn't even crack it," Kathryn exclaimed, turning it over and over in her hands. The egg was speckled and larger than any they'd ever seen.

Janet looked around nervously. "I guess we'd better put it back before the mother bird comes back."

But at that instant, a shrill call sounded from high above them. Looking up in alarm, they saw a bird circling and diving - a bird with a wing span so wide it seemed to block out the sun.

"Oh my gosh, that must be the mother bird!" Janet Lee exclaimed.

"Quick, put it back as fast as you can!" Kathryn urged.

Just as Janet Lee dangled over the edge, her sister clutching her legs firmly, the huge bird swooped down towards her head.

"Pull me up quick!" Janet yelled.

The girls huddled on the ledge, covering their heads with their arms, as the bird swooped and screamed and dove menacingly toward them.

"What'll we do? I've still got the egg," Janet Lee wailed.

Kathryn raised her head enough to recognize the bird as it flew close.

"It's a mountain eagle," she cried. "I've heard of eagles pouncing on lambs and sheep and chickens and flying away with them!"

"Could it carry me away?"

"Maybe, I don't know. But we've got to get out of here. Come on, run!"

Kathryn started to run, shouting and waving her arms, with her sister trailing behind. But after only a few steps Janet Lee fell flat onto the rocks, and the egg, still clutched in her small hands, smashed. Yellow stains covered her coat and gloves.

Desperately Kathryn pulled her sister up. Once again, the shrieking eagle dove straight for the children, talons outstretched.

"Lie flat!" yelled Kathryn. Both children dove for the ground, but Janet was a moment too slow. The mother bird snatched her hat in its claws and flew away with it.

* * *

It was a stricken Janet Lee that made her way down the mountainside, her face tearful, her coat egg-stained, her hat gone and her hair wild. Kathryn was shaken as well but regained her composure more quickly.

"You were the one that took that egg and broke it, so the mother eagle was really mad at you. And there was a baby eagle inside the egg, so you were the one that killed it."

Janet wailed forlornly, "But I didn't mean to!"

Back at the Governor's house, the girls had a hard time convincing their parents of the veracity of their story.

"You're telling us a mother eagle knocked you down, stole your hat, and threw an egg at you?" their father queried, eyebrows raised. "Are you sure she didn't just carry you home and drop you off in the front yard?"

"But it's true, Daddy! Except no, she didn't throw the egg at me. I dropped it," Janet said earnestly.

It took a hike back up the mountainside to convince their father. There, sure enough, were the shards of eggshell on the ground.

"Okay, girls," George sighed. "I just hope you don't dream about being chased by eagles tonight."

When the trio returned to the house, good news awaited at last. Permission from Copenhagen to proceed had finally come through, although they would not be allowed to follow George's original plans to cross the ice cap. They would have to fly the long route around the treacherous coast.

The White Bear was also departing the next day, carrying the expedition's gasoline ahead for refueling at appointed stops. Pete had received an excellent weather projection. The Sikorsky was in readiness, and Norman was satisfied with the film footage he had taken. The group decided to push off in the morning.

That evening, a farewell dinner was given aboard the ship by the officers, honoring the Hutchinsons as well as the Governor and his wife.

The Danish cuisine was excellent in some respects, if unusual in others. The purple-hued rhubarb soup was not exactly a favorite with the girls. They were intrigued when desert was passed around, however.

"These look just like gumdrops!" Janet Lee whispered to her sister, popping one into her mouth.

Kathryn followed suit, plucking up two of the little round balls that filled the dish and passing it on down the table.

At that moment, the captain inquired with a smile:

"And how are you liking the seal eyes?"

Sputtering, Kathryn spit the delicacies out into her napkin. It was too late for Janet Lee, who had already swallowed. She turned visibly pale. Even Blanche and George coughed, holding their napkins to their mouths.

And so passed their last night in civilization.

* * *

The next morning dawned cold and clear, and the whole crew was up early busy with the duties of packing, loading, refueling and making all ready for departure. While they worked on the plane, a string of native kayakers kept circling the vessel slowly. Interested Greenlanders watched their progress from the dock as well. When the refueling was finally complete, Pete and Joe went back to the radio station for the latest weather report, and George remained behind to try and warm up the cold, uncooperative motors.

Janet Lee and Kathryn, after saying goodbye to the village children, wandered over to the rabbit farm.

The Inuit who was in charge of the rabbits grinned at the appearance of the girls and wordlessly held out a smallish, wriggling gray creature to them. He pressed the rabbit into Janet Lee's arms, with a word that sounded like "Um-m." He thumped himself on his chest. Then he put his hands on the little girl's chest and repeated, "Um-m."

"I think that means he's giving it to us," whispered Kathryn, reaching out to stroke the velvety fur. "Do you remember the Danish word for *Thank You?*"

"Tek!" said Janet Lee, clutching the struggling rabbit close. The man beamed, and the girls beamed back, as they repeated "Tek, tek!" and left with their gift. What a wonderful memento of their visit to Greenland their new pet would be!

Halfway back to the dock, however, a thought struck Kathryn. She frowned.

"Remember what Daddy said about other countries not letting animals in? Remember what happened with Mike?"

Janet's face fell, but she rebounded quickly. "But this wouldn't be an American animal, it already lives here. And a rabbit doesn't meow like a kitten does. I don't think they make any sound at all."

Kathryn wasn't so sure, but soon the furry appeal of their new pet won out over her skepticism.

"Okay, then," she sighed. "But we're going to have to feed him. You run back to the kitchen and find some food for him. Bread, lettuce, anything. Wrap it up in your handkerchief and bring it back, quick."

Trying to avoid being seen by their father, who was trying to solve the cold engine problem, the girls snuck aboard the Sikorsky and stowed the rabbit in Mike's old quarters, the lounge seat. They made a soft nest for him with a pile of rags down in a corner, and were happy to see him attack their provisions with relish. The sisters took turns stroking the soft fur as the bunny nibbled away.

Suddenly there was a thump as a boat bumped against the hull of the plane, and voices and laughter reached their ears.

"It's Mother and Daddy! Quick, put the seat down!"

By the time their parents reached the cabin the two girls were sitting demurely atop the lounge seat.

"Well, it's a surprise to see you two ready and waiting! I thought we'd have to chase you down the way we usually do," their father said.

Both girls stared innocently up at their parents.

"What's that, Daddy?" asked Kathryn. George was carrying two heavy brown bottles.

"It's ether. I got it from the hospital, to pour on the spark plugs. That should make them start all right."

"Huh," said Janet Lee. "I thought ether was what they use to put people to sleep with for operations. Like when we had our tonsils out."

"You're right. And too much of it can kill a person, so stay away from it."

Carefully the men poured a few drops of ether into each cylinder in the motor, replaced the plugs, and tried once more to fire the engines. They backfired, belched black smoke, but thankfully began running.

All of Godthaab appeared to line the shores and the wharf, drawn by the sound of the motors, to wave and shout their farewells to their departing guests.

Preparing to cast off, George called back to Blanche. "Please put these two ether bottles inside the lounge seat, under that pile of rags."

Janet Lee and Kathryn jumped up on the seat again.

"What in the world did you do that for?" a perplexed Blanche queried. "You heard your father tell me to put these ether bottles away."

Thinking desperately, Kathryn protested, "But isn't ether dangerous? Maybe you should just throw it away."

"And I'm really sleepy," added Janet Lee, stretching out full length across the seat.

"Come on, girls, stop being silly and get off that seat."

Kathryn grabbed for the bottles. "Then let me help you, I'll do it!" Quickly she half-opened the lid, deposited the bottles, and slammed it shut.

The excitement of takeoff diverted everyone from further concern as the Amphibian accelerated and cleared the water. The plane leveled out and both girls yawned ostentatiously and curled up together on the lounge seat for a nap. Blanche, still puzzled, picked up her diary which she needed to update.

George flew low and close to the shoreline until a fog bank materialized. Climbing above it to four thousand feet, they flew for miles and miles with the fog beneath them like a blanket. Even after several hours no break in the fog could be found, so George swung inland over the mountains. From this high altitude they could spy the top of the famous Greenland ice cap, a magnificent sight glimpsed by only a few people at this time. An awesome mass of snow and ice had piled up over the centuries, forming a high point in the center of the continent.

As the adults in the expedition snapped photos out the windows, exclaiming at the scenes of natural beauty below, the children would occasionally be roused to sit up and look out. But they lay back down on the lounge seat quickly.

"George, do you think something's wrong with them?" asked Blanche worriedly. "They're so quiet! It's not like them."

Kathryn overheard them. "Don't worry," she whispered to her mother. "I think we're just having an off day."

Suddenly, a strong wind buffeted the ship and loose articles went scattering around the plane.

"It's Julianehaab, everybody!" Pete exclaimed, pointing directly ahead.

And there, at the very southern tip of Greenland, lay their destination. Julianehaab was another little township at the base of a rocky hillside, ringed by a natural horseshoe-shaped harbor.

Everyone was glued to the windows watching the descent when Blanche sniffed loudly.

"George, do you smell something funny?"

Kathryn and Janet Lee had already jumped off the lounge seat.

"It's here, right by the lounge. It smells like ether!"

"Oh no, maybe the bottles have broken," cried her Kathryn.

Both girls scrambled to lift the seat top. Now everyone was smelling the pungent odor.

"Better get those bottles and rags out of there," called George.

Kathryn plunged her arms inside the compartment and lifted out the rabbit. Eyes closed, limp and lifeless, the furry bundle dangled from her hands.

"He's dead!" wailed Janet Lee, already beginning to tear up.

Both girls were so upset by their short-lived pet's evident demise that their parents were hard put to scold them for harboring a stowaway.

As they were landing in the harbor, however, the bunny showed signs of incipient life. His whiskers twitched feebly.

The girls begged their father to hurry and land. Although a welcoming committee awaited them, Governor Isben upon understanding the crisis hustled the family through the crowd and on to the hospital with their limp bundle. But it was too late. The little rabbit was officially pronounced dead.

For the first time the joy of arrival was marred by sadness as once again friendly Greenlanders crowded around to greet the expedition.

Kathryn, Janet Lee and their father climbed up a hill later that afternoon. The view of the harbor was breathtaking as they took turns with a shovel. The cloth-wrapped bunny was carefully placed inside the hole and Janet mournfully placed over it the headstone George had painted and lettered for them earlier. It showed an American and a Danish flag, and this inscription:

In Memory of Ether, Our Rabbit
September 7, 1932

* * *

Julianehaab, with a population of some seven hundred souls, was larger than Godthaab. More boats entered the harbor here than at any other port in Greenland. The climate was milder, also, at this southernmost point. The only trees found in all of Greenland grew nearby, where a forest of birches lined the Tassemint Fjord.

As before, the Hutchinsons were quartered at the Governor's comfortable home and enjoyed good Danish cooking (though everyone was on the lookout for seal eyes). Fresh produce from the garden was most welcome. Nights were nippy, however, as only the first floor of the house was heated, and washing up in the icy water drew shivers.

The weather turned bad soon after their arrival, with only a few breaks in the stormy conditions allowing for more film footage to be shot. The Danish warship from Godthaab, (which had been delayed for a day because of the terrific storms), finally showed up with the fuel, and the men immediately filled the Sikorsky's gasoline tanks. Now only the weather delayed their departure for Angmagsalik on the east coast of Greenland.

The boredom of waiting was offset by card games in the evening. Sometimes the Governor's daughter Elsa, who had a beautiful voice, would sing. In the daytime the village was further explored, weather permitting. The girls loved riding on Elsa's tiny Icelandic horse. Wherever the girls went, curious Inuit children followed behind like eager puppies.

The small native houses were colorfully painted in red, yellow or green. Tiny garden plots were surrounded with white fences. The town was frozen in for almost six months of the year, with an average winter temperature of five degrees below zero. In the summers, it could warm up to sixty degrees.

The Hutchinsons observed that the native women seemed to do all the hard work. Even when it was raining hard they could be seen unloading heavy lumber from boats in the harbor.

In addition to goats and sheep, the settlement boasted a large rabbit farm. Strangely, the girls seemed to have lost their enthusiastic interest in rabbits and declined to visit.

* * *

Late in the evening of the fourth day, Thane the radio operator reported a favorable weather forecast at last. Sure enough, that night the stars once more brilliantly pierced the inky sky and the moon brightened the craggy landscape with an unearthly beauty. Accordingly, George planned an early takeoff for the morning.

Everyone was up by 4:00 AM to make final preparations. George had to wait for the latest weather report, which proved to be less than a hundred percent favorable after all. Elsa had prepared sandwiches and fruit for them to take along, and the Governor sent a thirty-eight-pound fully dressed lamb to be presented to the Governor of Angmasalik as a gift.

It was a challenge to start the cold engines once again, but by seven o'clock both motors were purring steadily, all were aboard, and of course the entire town was awake and down by the wharf to watch the plane take off from the harbor.

The White Bear was departing as well, officers and crew lined in formation on deck to bid farewell. The girls thrilled to see the ship's cannon belch fire in final salute to Julianehaab and to the Flying Family, as the Sikorsky went skimming along the water alongside the huge warship.

This ascent, deterred by rough swells and an extra heavy load, was more difficult than usual. Finally the *City of Richmond* kicked clear of the last wave and was airborne once again.

13

THE WORLD IS LOST

September 11, 1932.

The expedition's flight path now lay due south and then east, around the very tip of Greenland. The sixty-mile stretch from Julianehaab to the Lindenwald Fjord on the east coast traversed some of the worst flying country in the world. George climbed to eleven thousand feet to get above the heavy winds and mountains. The temperature dropped to ten degrees, and it became difficult to stay warm in the cabin. Blanche and the girls put on extra wraps.

Even at this altitude, the plane was barely clearing the tops of the highest ranges. The group looked down upon a barren, rugged landscape bisected by deep, winding fjords packed with ice. The sight was incredibly beautiful, yet harrowing.

George breathed a sigh of relief as he cleared the mountains at last and turned northward at the Lindenwald Fjord. Over land, a safe landing in the Amphibian would be impossible. Over water, the whole ocean was a potential landing field. Descending to three thousand feet, George and the others were still exclaiming over the beautiful scenery. The air was clear and the sky impossibly blue. Cameras were rolling. These would be the first motion pictures of Greenland ever taken from the air. The temperature rose to thirty-two degrees. As they flew northward, the icebergs became more numerous, dotting the ocean below like floating diamonds.

For two hours all was peaceful. The Sikorsky was making excellent time, helped along by a strong tailwind. Blanche was writing in her diary, Kathryn was coloring in a picture of the coastline she had drawn, and Janet Lee was fast asleep. Pete, by

George's side, was paying close attention to his maps. Jerry was frustrated that he received no station replies, though he radioed their position every fifteen minutes.

Suddenly a large flock of geese materialized in front of the wind screen, flying straight toward the engines. The big birds veered sharply in front of the plane, and George immediately banked to the left. Too late. The next sound was a sickening thud. One of the big birds crumpled and fell toward the water.

The top left wing was dented and a patch of the fabric torn away. Ship and bird had collided with great force. The damage to the plane was not serious, but George, like most pilots, was superstitious. Hitting a bird in flight had always been construed as an omen of bad luck.

Not more than ten minutes later, low lying clouds heavy with snow began appearing ahead of them, apparently forming out of nothing. From this point on, everything started to go wrong.

* * *

George was afraid to fly above the clouds because it would reduce visibility. They were nearing Angmagsalik, which Pete calculated to lie some seventy miles to the north. With zero visibility, they could easily miss it on this intricately carved coast. Flying under the clouds seemed the most prudent course, so George descended to an altitude of only 300 feet above the water. Immediately, the wings began to take on ice.

Once again Jerry tried to signal Angmagsalik to find out if it was clear there. Still no response. Doggedly, he kept trying.

Things were getting more tense by the moment. The Sikorsky was clearing the icebergs by only a few feet as the craft skimmed along just above the water. George had to carefully maneuver around and over the huge floating barriers, constantly making sharp, banking turns. It was harrowing. He tried once more to climb up through the clouds, but the wing ice began accumulating much faster. They would have to stay low and pray.

* * *

At about this point in the crisis, the account that George Hutchinson gave of the ensuing events in his book *The Flying Family in Greenland* and elsewhere begins to diverge sharply in some areas from that given by the two living eyewitnesses, his daughters. Only 42 out of the 297 pages of his book are devoted to the Greenland adventure. Yet, it is the single event that brought the Flying Family international recognition and fame. So, why did the hero of the tale dwell so briefly on the telling?

Janet and Kathryn felt it was because the failure of the undertaking was so very painful to their father. It hurt his pride, it hurt his pocketbook (the loss of the Sikorsky and the mission itself meant the loss of a huge investment) and, should he allow himself to grant any credence to the accusations of his critics, it would hurt his self-image. George Hutchinson needed that self-image intact. Thus, he told his version of the crash and rescue, leaving out certain points and emphasizing others, and he told it briefly.

* * *

Jerry, desperately trying to radio an SOS, finally picked up the Scottish fishing trawler *The Lord Talbot*, twenty-four miles away. "I've picked up a ship!" he shouted. "Everybody stand by!" Scribbling furiously on a pad, he began sending and receiving.

He turned back to the tense group. "They're about two hours away and can pick us up if we can manage a landing. They say it's snowing where they are, too."

The snow was freezing on the Sikorsky's windshield now and visibility diminishing by the minute.

Here, George's retelling of events begins to sharply diverge from remembered accounts by his family. According to George, he simply evaluated to situation, realizing he would soon be flying blind, and decided to land on the water and taxi toward Angmagsalik. He asked Jerry to notify T*he Lord Talbot* of their plans and ask them to come as quickly as possible.

Seeking out the lee side of an iceberg, he calmly maneuvered the huge plane among the loose, floating ice and set down on the rough sea. George claimed that no one was showing any signs of fear.

* * *

Kathryn remembered the progression of events differently, and vividly.

She wasn't really frightened until the plane was flying so low they were just clearing the waves. Sometimes they actually skimmed the top of a particularly rough swell. Looking out, she could see the ice on the wings, weighing them down toward the menacing sea. In those days there were no de-icers. The force of the snow and wind increased to blizzard strength.

Still, it was the tension between the adults that finally awakened actual fear. Both girls had been in precarious situations before, situations that had ultimately turned into rousing adventures. Neither girl doubted that their father could get them out of anything. But it was becoming apparent, to Kathryn at least, that the rest of the crew was rapidly losing confidence in George Hutchinson's ability to pull this one off.

Janet, roused from sleep at last, was aware of commotion but not really of danger. Blanche hustled the girls to the very back of the cabin, as far away as possible from the adults.

The "boys" gathered anxiously around their pilot, and as their anxiety rose so did their voices. The crew wanted George to go on and land immediately. George held out for forging ahead toward Angmagsalik, which, as it turned out, was only thirty-eight miles ahead. They might have made it, although poor visibility, ice on the wings, and the looming threat of the icebergs all made it perilous. But then, so was the prospect of landing in rough, iceberg-flecked seas. Some of the waves reached heights of twenty feet. Pieces of icebergs constantly broke off. The center of gravity shifted in the ice and the huge monoliths could turn over suddenly and dangerously.

* * *

Strident voices arose from behind the cockpit.

"You've got to land, damn it! You're crazy to think we can make it!"

"Colonel, *The Lord Talbot*'s on its way. We'll be found. We're going to crash anyway if you go on!"

Only silence from their pilot.

"God damn it, there's four of us and one of you, you've got to listen to us. Don't you care about what happens to your kids?"

But his crew's alarm and anger did nothing to deter George Hutchinson.

"I'm the captain!" he yelled back. "I think we can make it. We've got to be getting close to the town."

Desperately, Blanche tried to get Kathryn and Janet Lee to color, to play a game, anything to keep them from focusing on the reality of their danger and the rising conflict between the men. But Kathryn, at least, could not block it out.

When one of the panicking crew members actually lost control of his bladder, anxiety sharpened. Perhaps, at last, there was good reason to be afraid.

Through it all Blanche was a center of calm and strength in a whirlwind, managing her own fear by attending to her daughters and trying to reassure them.

The crisis peaked when one of the men held a heavy monkey wrench against the side of George's head.

"What do you think you're doing?" he yelled back at the crew.

"You won't listen any other way. We're not all going to die because of you. Bring this plane down now!"

Without a real choice, George focused on the ice-flecked sea before him. Snow swirled relentlessly as he desperately tried to find a clear, open space for landing. Lower and lower the Sikorsky flew, as the waves began to plop against the hull. The "flying yacht" was hitting the top of the water in a wild ride like a bouncing speedboat.

Blanche had been praying steadily, both to herself and aloud with the children and sometimes the men. It would be so easy for a wing to clip an iceberg!

* * *

The Sikorsky touched down at last, tossed around by the turbulent sea like a child's toy but miraculously still upright. The danger, however, was far from over. Towering waves threatened to engulf the plane. Huge icebergs loomed. The crew of eight were down, and alone, in the middle of a snowstorm in the wastes of Greenland. At least they had managed to notify the Scottish trawler. Jerry continued to send out the relief call SOS until the radio batteries became waterlogged and gradually died out.

George began an attempt to taxi in the general direction of Angmagsalik, but their progress was fraught with danger. There was the possibility of being capsized by the treacherous shift of an iceberg. To maneuver around the looming monoliths they had to open and close the windows in order to see. Freezing seawater was thrown into the cabin.

With a sickening crunch, a big piece of ice that had broken off from an iceberg hit the hull. The plane began to take on significant water, and George realized that they could sink. Worse, the impact from the floating ice had damaged a fuel tank, and precious fuel was leaking from the craft. George realized he would be lucky to find shore, let alone their destination. He began to taxi as fast as possible toward an outcropping of ice-covered rock barely visible ahead in the swirling snow. In the space of what seemed like a few minutes, everything had come down to a matter of survival.

* * *

Elsewhere, the fate of the Flying Family was being followed anxiously.

F. Spencer Chapman's retelling of the adventures of four British explorers in Greenland, *Watkins' Last Expedition,* details the events surrounding the rescue of the Hutchinson expedition. Chapman and two others (the fourth member and leader of their expedition, Gino Watkins, having been killed just weeks before while seal hunting) arrived in Angmagsalik shortly before the Hutchinsons were expected. When they went to the Governor's house to occupy their usual quarters there, they were told that they were reserved for "The Flying Family." The explorers, on their ship *The Stella Polaris*, had been in touch with *The Lord Talbot* as that crew conveyed messages of sympathy on the loss of their comrade. On September 11 Chapman and crew observed the trawler, heading for port in Angmagsalik, turn suddenly south and make off at great speed.

In *Watkins' Last Expedition* Chapman related the reaction of their crew to the news of the Hutchinsons' crash and the search that was underway. When they reached Angmagsalik at about 4 PM that day, everyone in the settlement was excited and concerned. Stilling Berg, the weatherman, had continued to send weather reports of the snow and unsuitable flying conditions. About midday he heard a distress call from the Hutchinson expedition stating that they would have to make a forced landing. Their SOS gradually faded out over the next hour or so. *The Lord Talbot* too had picked up the message and rushed to the rescue.

* * *

As the group of survivors approached the island dead ahead, the ominous hiss of water below in the hull let them know it was not a moment too soon. The island itself had no beach, only jutting rock cliffs and big boulders. George was forced to run the hull onto the rocks to secure the plane while the passengers jumped ashore. The anchor could not hold here, because of the sloping rock and the depth of the water.

Janet and Kathryn were lifted out of the sinking plane by the men and passed along hand to hand to the shore. Then, working as fast as possible, the men applied themselves to unloading everything that might be of use from the Sikorsky. Food, clothing, blankets, cushions, guns, Sterno, oil, tools, instruments, radio, and more were piled up on the rocks. Blanche and the girls helped shift everything to higher ground away from the water.

Waves and floating ice battered the plane unmercifully, even as the crew labored to empty her. In despair George looked out at the open, angry sea, wondering how a rescue vessel could possibly see them through that stormy field of floating ice. Two huge icebergs were slowly drifting closer. George felt the Sikorsky shudder and lurch as he, the last one aboard, jumped ashore.

* * *

Life magazine's cover of the year for 1932 would feature the photograph that dramatically captured what came next. Silhouetted in stark black and white, Blanche, Kathryn and Janet Lee stood atop a huge boulder, watching the Sikorsky falter and sink beneath the icy Arctic waters.

George Hutchinson stood apart and watched not only his cherished Flying Yacht but his dream of making history sink as well. For the only time she can remember, Janet Lee saw her father weep.

Along with his grief, there was anger as George thought of the years of hard work and expense to prepare for this expedition. Bitterly, he blamed the bureaucratic shortsightedness that had sent them on the treacherous course that resulted in the crash.

Perhaps in this way George protected himself from possible guilt or self-doubt. Disregarding the fact that he had chosen to take off that morning after adverse weather warnings, he could focus instead on the wrongheadedness of the Danish government. As his plane sank he still managed to keep afloat the image of himself as imminently capable and triumphant in the face of odds that would have daunted ordinary mortals.

Little time remained for emotion, however. There was too much to do, and the activity helped keep the castaways warmer and rally their spirits as well. Jerry discovered that the wireless had gotten wet in transport. While it could still receive, it could no longer transmit. Everyone was relieved to hear that several boats, including *The Stella Polaris* and *The Lord Talbot,* were speeding to the rescue. Yet how frustrating to know that there was no way to direct the would-be rescuers to their true location, which of course lay some distance away from the spot from which the first S.O.S. was broadcast.

* * *

Working as swiftly as they could in the cold, the little band shifted everything from the shore to a spot high up on the side of the cliff, about two hundred feet above the water. Here, a three-walled natural formation of rock provided a type of shelter. Everyone worked at putting supplies in order.

A survey of the terrain revealed that they were on an island of rock, in one of the deep fjords that scored Greenland's coast. Pete and Joe climbed to the highest jutting boulders and placed white cloths as distress signals.

The rugs and cushions from the plane were laid on the floor of the shelter. A rock fireplace of sorts was fashioned in one corner, and a fire of Sterno, grease, and oil was lit.

Blanche inventoried the food supply anxiously. "Thank goodness for the lamb," she said. "It should last us a good two weeks." The thirty-eight pound carcass would prove to be truly worth its weight in gold to the hungry castaways.

Everyone put on what extra clothing they could. They weren't outfitted with Arctic gear, but their flying togs and heavy coats kept them surprisingly comfortable, except at night.

Precious flares, guns, ammunition were secured in a dry spot. Norman stored his camera equipment and film under some rocks. Jerry kept working hard on the radio, hoping to repair it. Janet Lee and Kathryn helped their father select rocks and stack them to form a fourth wall to the shelter. They also built up the other walls.

George knew they needed a roof to make their shelter secure against the elements. One wing of the Sikorsky was still sticking up out of the water. Climbing out precariously on the rocks, he finally managed to salvage a large piece of canvas from the wing. The men fashioned it into a roof for the shelter, and another piece made a door. Now they would at least be able to keep out the snow and wet.

It was about thirty degrees, and still snowing. Scouting around for water, they discovered streams of melting ice flowing down the hillside and collecting in depressions in the rock.

The little band took stock of their situation. Even after the food supply was depleted, they could catch fish and shoot birds. There was drinking water and clothing at least warm enough to keep from freezing. The fuel supply was the real worry. How long would the Sterno, grease and oil last? There was nothing to burn on the barren island, no signs of vegetation.

Each adult stood watch at two-hour intervals, scanning the sea with powerful binoculars for rescue boats. Even the girls stood their turn, but only for an hour each and with a watchful parent nearby. This helped them feel they were doing their part. With the snow still falling, it seemed impossible that anyone would see them or be able to venture close enough through the ice field to reach them.

Now that they were "safe" on the island with their improvised shelter, Janet Lee and Kathryn looked on their predicament as something of an adventure. The adults showed no outward signs of worry. As George's account stated, "Nobody whined." Blanche, with Norman's help, prepared a lamb stew that all pronounced delicious. As they ate, joking about the Governor's gift, the sound of icebergs colliding, splitting or turning over in the water echoed all around them.

"It sounds just like cannons," Janet Lee declared.

The sky finally cleared and the bright stars appeared. Before turning in, each person drank a little brandy to help warm their insides. The group slept huddled spoon-fashion on the cushioned floor. Blanche and the girls were in the center, with the men on the outside. Everything they could get their hands on was piled on top, and they managed to keep fairly comfortable through the cold night.

The vigil went on. Every two hours the one on watch would return to camp and a new sentry would take his place. Twice the first night, the cry of "A ship! A ship" roused everyone from bed. Rushing out into the night they peered vainly through the darkness. Disappointment was heavy when they realized that the light was only blue reflections from the icebergs, as the moon and stars shone down in their brightness.

And so the first long night passed.

* * *

News of the Flying Family's disaster had reached the ears of the press, and of friends and family, in the world they had left behind. From the time they had flashed their last SOS, the concern of the nation and indeed the world turned toward the fate of the Hutchinson expedition. Headlines screamed: *"The Flying Family is Lost!" "A Mother and Two Children Stranded on the Ice!" "Eight People Missing Among the Treacherous ice Floes of Greenland!" "All Feared Lost!"*

Well-known radio commentator Lowell Thomas reported that rescuers had searched so long with no results that it was believed the Hutchinsons and party had sunk with their plane.

The families of both George and Blanche were deluged with press attention and the calls of people who drove them crazy with their concern and questions. Family members took to walking the streets to keep away from the bombardment. Yet of course they were eaten up with the ultimate question: *Can they possibly still be alive?*

* * *

The bright sun rising early next morning restored hope to the castaways. After a liquid breakfast of coffee, hot chocolate, and soup, everyone got busy.

Jerry started building a crude radio receiver, incorporating salvaged equipment from the plane. George took his daughters on an exploratory hike to survey the island. It was hours before they reached the top of the rock mass, a high point from which they could at last scan the horizon. Spirits fell at the unbroken view of icebergs and drifting ice. George could just make out the fjord that led into Angmagsalik Bay. From their vantage point it looked only a short distance away, yet that distance meant the difference between the success and failure of his plans.

Janet Lee and Kathryn had a good time mountain climbing and playing explorer. The seriousness of their situation did not sink through to the children. They felt safe, simply immersed in another adventure.

Lunch back at the camp was boned chicken, green peas and canned prunes for dessert. Norman had been making more movies, and as he was telling the group about some scenes he had shot that morning, Jerry, still working on the radio, suddenly shouted.

"I've fixed it! It's working!"

Everyone gathered round. "Will we be able to radio our position?" asked Blanche.

"No. But we'll be able to hear what's going on. At least we'll know when they're close."

As soon as he had attached the final piece of equipment, the aerial, a series of dots and dashes flashed over and over through Jerry's earphones. He began to jot down notes.

The company pressed anxiously around him, waiting. His face was troubled as he looked up from his pad.

"It's not all good news. I heard *The Lord Talbott* talking to the station at Angmagsalik. They're still looking for us, but they're asking permission to abandon the search. They circled the spot we sent our SOS from and didn't find anything."

The stricken faces around him said it all. Then suddenly, "Wait! There's more." Furiously he began scribbling again. After a couple more stops and starts, he turned to the group.

"Angmagsalik talked to *The White Bear* who told them to instruct *The Lord Talbot* to continue searching at least another twenty-four hours. The ship radioed back agreement. And there's more help coming."

Everyone hung on his next words: "Two Britishers, Chapman and Riley, are following the coast southward in their motor launch looking for us. The Scottish trawler signaled another ship by lights - it has no radio - and it's searching, too. Then there's another fishing trawler, the Danish ship *Swordfisken,* and another Danish patrol boat. Plus, the signals from *The Lord Talbot* are really strong now. They must be getting close."

Relief and laughter followed his good news. Surely, with all the searchers, they would be found.

* * *

Now the watch for the hoped-for rescuers began in earnest. Climbing out on the rocks in front of the camp, the group passed the field glasses around hand to hand. Eyes grew weary from squinting out over the horizon. Occasionally they were cheered by Jerry's update that the *Talbot*'s signal was getting closer. Still, the time passed slowly.

As the sun sank, the icebergs seemed to take on the form of fantastic, grotesque figures. Some of them looked like ships. Once, all eight were so convinced that an iceberg actually was a boat, they wasted several of the precious flares and a good number of ammunition rounds trying to attract its attention.

Although it would be easier, George reasoned, to sight a ship at night, keeping a light going by burning precious fuel was out of the question.

Jerry's face was serious as he reported the latest news. *The Lord Talbot* had not received the original position of the crash clearly, and was now turning back northward toward Angmagsalik, following the coast.

Still the watch in the night continued. Darkness and silence surrounded them, except for the occasional cannon-like report of an iceberg breaking in two. Finally, Jerry joined the group on the rock.

He spread his hands in a helpless gesture. "It's gone, folks. The batteries and the tubes both went haywire."

That meant no more radio, no more news. Silence and cold seemed to penetrate even deeper into their very bones.

Blanche put the girls to bed after preparing some hot beef broth. As she tucked the heavy rugs and blankets around them, Janet Lee looked up at her mother.

"Are we really and truly lost?"

"I do believe we are. For now."

"But *we* know exactly where we are, right?"

Blanche smiled. "That's right, dear."

"Then we're not lost," Janet insisted, snuggling down into the covers. "It's the *world* that's lost!"

* * *

Out in that larger world, the Hutchinson expedition was indeed feared lost. Nobody, however, wanted to give up on even a slim possibility of rescuing a family and two little girls from an icy Arctic death. Everything possible was being done to exhaust every search possibility, however dubious.

The Danish Newspaper *The Times* reported on Tuesday, September 13:

"COPENHAGEN, September 12. The Danish Inspection ship Hvidbjorn stationed on the coast of East Greenland sent the following message to-day: -The Lord Talbot reported 7 a.m. That the search had given no results.

A Danish seaplane of the Knud Rasmussen expedition, piloted by Lieutenant Rasmussen, left Julianehaab at 12 noon to follow the coast to Angmagssalik. Although Mr. Hutchinson had no permission to land in Greenland and was warned not to do so, everything will now be done to find him. The chance is considered very small, but he may have reached one of the small rocky islands or be on an ice floe...

A message from Berlin says that Universal Pictures Limited has ordered the pilot Herr Udet, the War "ace," who is in Greenland with a film expedition, to take part in the search with his three aeroplanes."

* * *

George drew the lot for the first watch of the night, but the men decided to make it for one hour only. Each person on watch needed to be as alert as possible. It was a bitterly cold, clear night. As he gazed out over the water, George's thoughts returned to the reasons for their predicament. Again he placed the blame fully on the Danish government and congratulated himself for handling the crisis.

It was important to George that this setback not be allowed to stop him from fulfilling his plans and dreams for the Flying Family. Adverse circumstances, even those as extreme as the challenge that faced them now, brought out his competetive, fighting spirit. In his mind he compared what the Flying Family was setting out to accomplish with aviation to the challenges faced by the first American pioneers.

Jerry replaced George at watch a little after one o'clock. Soon thereafter a loud shout woke everyone.

"A ship! It's a ship, no mistake this time! Come on, everybody!"

As the group began to stir and struggle down to the vantage point, Joe muttered, "If he's wrong again, please help me dunk him in the ocean."

But when they had scrambled onto the rocks, Jerry pointed triumphantly. "Two lights! See, right over that iceberg?"

George squinted. Eyes unaccustomed to darkness had a hard time making anything out.

"I don't see anything," said Kathryn anxiously.

But one by one each person was able to make out what looked like two lights moving on a boat, one at the bow and one on the stern. "I see it!" "Me too!"

A third and brighter light, high above the other two, began zigzagging slowly from side to side. "They're using a searchlight," cried Jerry.

There were only two flares left. In a second Pete had lit one, and the blaze illuminated the landscape with an eerie red glow.

Norman rushed down into the camp for his film. He was prepared to sacrifice several rolls if necessary, to build a signal fire. Joe set fire to a small can of gasoline, and George ignited a can of grease.

Tense minutes passed. There was no sign of recognition from the boat.

Sighing, Norman turned to the pile composed of film rolls. The documentation of their expedition, groundbreaking footage of exotic terrain and peoples, would have to be sacrificed on this last hope. As he lit the pyre it burst into a great orange pillar of fire. Hopes rose like the flames shooting upward. Surely with this bright beacon, plus the flare which continued to burn, they would get the attention of the ship at last.

All the company stood poised on a knifeblade of tension, straining to see, waiting. Would it all be for nothing?

14

THE LORD TALBOT

Finally the center light stopped its back-and-forth movement. Then, with excruciating slowness, it began to blink, blink, blink.

"They see us!" Jerry cried. "Those are signals they're flashing!"

Unfortunately, try as they might, no one among them could decipher the code message the ship was sending. The reflections from the light rays as they struck the water and the icebergs were too distorting.

"Guns!" George shouted, "Let's shoot off the guns!"

As Jerry and George took turns firing the guns, the lights were still too far away for the group to know that the ship had recognized the retorts. But evidently the ship had turned broadside, for now only one light was visible, and now it started to move slowly straight ahead toward the island. Now Jerry, looking through the binoculars, was able to make out a few words of the signal.

"They keep repeating something about *Chapman...Chapman and Reilly*," Jerry puzzled. "Wait, here it is: they're asking *Are you Chapman and Reilly?*"

The Lord Talbot had been informed that the members of the Watkins expedition, Chapman and Reilly, were also in the vicinity in their ship *The Stella Polaris*, searching for the Hutchinsons. The *Talbot* Captain needed to determine if the signals were coming from other rescuers or from the crash victims themselves.

"They think we're someone else," Jerry called back. "Wait, here comes more."

But it was too foggy, and the refracted light too distracting, to decipher the flashed code exactly.

George called, "Go on and set off the other flare, Pete!" How frustrating that they had depleted their store of flares on ghost ships.

The first flare burned out. Pete immediately lit the second and the camp lit up brightly.

* * *

Unbeknownst to the castaways, *The Lord Talbott* had been repeatedly signaling:

If Hutchinson, light two fires.

Now, unbelievably, the ship was turning, slowly moving away in the opposite direction. The crew had seen only one signal fire, interpreting that to mean that they had succeeded only in contacting another search party and not the Hutchinsons.

This couldn't be happening.

In sheer desperation, the men piled two more rolls of hundred-foot film on top of the remaining small grease fire. With this conflagration, all the priceless film footage of the entire expedition was destroyed.

But it worked. The film burst out into bright flame for a few short minutes. Pete waved the flare around and around. They had lit two fires after all, by accident, luck or providence.

* * *

Everyone held their breath. Would the ship see?

It did. With excruciating slowness, the *Talbot* halted in its departure. Once again Jerry began to call out the letters as they were flashed in code by the searchlight:

We are coming. Be patient.

Floodgates of joy and relief broke over the little band on the shore. Holding hands, dancing, hugging, cheering, they celebrated as they watched the ship wend its way ever closer through the intricate maze of icebergs. The last flare soon burned out, but they still had several flashlights, and twirled and flashed them repeatedly so that the ship's crew could pinpoint their position. The wait was nearly over.

* * *

The Scottish vessel was ill suited for working in ice, and it was getting dark. However, Captain and crew had no idea what the condition of those on shore might be, so they pressed on with all speed. It was harrowing to crash through the ice at a speed up to 11 knots. After three hours of desperate work, they reached the shore about 10:30 P.M. The swell was too great to launch a rescue party in the dark, so they were forced to wait for daylight.

With salvation so close, everyone's appetite reawakened and in the five hours it took the rescue boat to come within boarding distance the group hungrily dispatched everything in the camp.

With the dawn, a rowboat put out from the ship. Two strong crewmen men rowed swiftly to the water's edge where the group waited.

A third man, standing in the bow, called out in a thick Scottish brogue:

"Everybody all right there?"

* * *

And thus, aboard the fishing trawler *The Lord Talbot* out of Aberdeen, the voyagers found themselves safe at last. For the past forty-eight hours the crew had abandoned their trade, which was fishing for halibut, to devote themselves to the search. (Later, back in port, the ship's owners would, unbelievably, dock the pay of these selfless seamen for lost time and income. The Captain would be fired.)

As the girls were lifted aboard by the rough but gentle hands of the fishermen, they were greeted by Captain Thomas Watson. Described by George Hutchinson as "...about the finest and most big-hearted man I have ever met in all my life," the group were to get to know the Captain well.

Skipper Watson told the story of his crew's part in the rescue many times in the days to come. In one newspaper account he related:

After nineteen hours' searching for the missing plane a fire was seen about ten miles off. This was at ten o'clock on Monday night. Then a flare went up - the last, I learned later, and I have the handle of it yet. It will be a memento I will hang on to all my life.

In our anxiety to go to the aid of the family we set off full steam ahead, but came to a sudden stop after fifteen minutes. We had struck an iceberg. Fortunately, no damage was done, but it was a warning to slack off and go gingerly.

It was an eerie vigil. The night was exceptionally dark, and lighted up only by our occasional rockets. For three hours we crept along in the icepacks and among the icebergs. We will never forget the experience.

The mate was up in the crow's nest, two men were in the bow, and there were two on either side warding off the ice with the oars from the small boat. There was also a man at the echometer watching the water depths, and I had another with me on the bridge. My hand was never off the engine-room signal.

So dangerous was it that we signalled the letter "W" to the Star of Victory and Mount Ard, other two Aberdeen vessels helping in the search. This was to warn them that without a searchlight progress was impossible.

Another member of the crew added to the description of the extreme danger the ship's crew faced in the rescue attempt:

We knew that we were heading through a dangerous area of iceberg. Even from below we could see them towering up - almost as high as the Aberdeen Town House. We had to keep dodging them in the dark, depending on our searchlights to pick out a clear course...Every command was obeyed on the instant. Every member of the crew at his particular post realised that it was a matter of life or death, and so all of us did what we possibly could to bring off the rescue of the stranded fliers.

* * *

Retracing its path through the treacherous ice fields, The Lord Talbot headed for Angmagsalik. On the way the Hutchinson party learned just how close a call their rescue had truly been. The ship had been on its way back to the town to abandon the search when the mate in the crow's nest first sighted their fire. And, of course, they learned about the fortuitousness of lighting the second fire; had they not, they would probably have been abandoned there. Without fuel, they would have perished.

To warm their chilled guests, the crew passed around shots of brandy. Janet remembered this as her first "real drink." The Captain and his crew were hospitality itself. The ship finally reached Angmagsalik at three o'clock in the afternoon.

At the village, the rescued and the rescuers were met by the Governor, the Parson, Mr. Stilling Berg (the wireless operator who had broadcast the messages received from the Hutchinson plane,) and a colony of Inuit.

The members of the Watkins expedition, Chapman and Riley, met the Hutchinson party at the village as well. Chapman stressed George Hutchinson's extraordinary luck in spite of the loss of his aircraft and remarked on how "little cast down" the family was by their ordeal. He also commented on the stir created among the Inuit by Blanche's rouge and smartly cut jodphurs.

Chapman was not pleased, however, to learn that the newspaper *The Daily Herald* had chartered the *Talbot* to take the Hutchinsons back to Scotland. The little crew had hoped for a lift back to Lake Fjord on the fishing vessel themselves.

Janet was suffering from a bad cold, but the brandy helped. In the village she was presented with a handcarved native doll by the villagers when they learned that her own favorite had been washed away in the crash. She had cried bitterly at its loss. She christened the new addition "Eski" in their honor.

It didn't take long for the good news to be spread from the ship's radio to the world at large. The dramatic rescue of the Flying Family was huge news both at home in America and abroad. Prayer vigils for the missing Hutchinsons and crew, and especially for the little girls, had been taking place since their loss was reported. Radio broadcasters shouted the news, and the newspapers once again carried dramatic headlines:

"Thrilling Rescue of Flying Family!" "Lord Talbot Rescues Flying Family!"

* * *

Skipper Thomas Watson and his fifteen crew members were veterans of the harsh conditions they routinely encountered as they ranged far from their homeland into Arctic waters in search of some of the world's best fishing. Indeed, Watson was known for being the only Aberdeen skipper to brave those far-off waters throughout the whole year including the severe Arctic winter. Captain and crew often had to break up the ice from the vessel's deck with iron bars in the morning before fishing operations could proceed.

These men seemed remarkable to the Hutchinsons for their cheerfulness and extreme hardiness. The voyage back to Scotland was a physical ordeal, however. Bad weather and high seas accompanied the voyagers on their return trip. The ship, although large enough and well-equipped for a fishing trawler, was still small enough to roll badly in rough seas.

Janet Lee and most of the others suffered from severe seasickness. The seas were so rough that even some of the seasoned sailors fell ill. A little good Scots whiskey did seem to settle the stomach.

When the Flying Family and crew finally reached Aberdeen about 1 P.M. on September 21st, they were greeted with waves of warmth as crowds of enthusiastic Scots lined the docks to cheer their arrival. They were paraded through the packed streets in an elegant open horse-drawn carriage, waving to the cheering onlookers. Later, the Lord Provost presented them with a key to the city.

The local newspaper reported on the family's arrival at the Sutherland Arms Hotel, where Blanche described her experience of the crash and rescue in emotional terms.

"It was a terrible experience - one which I shall never forget, she said. It was bitterly cold when we were eventually put ashore. We could not get a fire at all because there was absolutely no wood. We sent the children running up and down the cliffs with the few things we were able to salvage to keep themselves warm. The things were very few, because the amphibian sank twenty minutes after we got away.

The kiddies managed to keep warm, but their feet were cut - in fact, they were bleeding. After that I made them skipping ropes, and they kept warm by skipping.

Mrs. Hutchinson shuddered as she spoke of the voyage home in the Lord Talbot.

We were all sick, she said, for nearly five days. Janet was the best sailor. Captain Watson and the crew were wonderfully kind. "

* * *

The London newspaper *The Daily Herald* arranged for a chartered plane to fly family and crew to London. The Lord Mayor himself met with the Hutchinsons and presented them with the key to the city.

Joseph Kennedy, then U.S. Ambassador to Britain, invited the Hutchinsons to the American Embassy for a State dinner. Politely, George declined. A staunch teetotaler, he didn't approve of

Ambassador Kennedy's alleged connections to rum running during Prohibition.

The family were given an escorted tour of the city and all the historical attractions such as Big Ben, Parliament, and the Changing of the Guard at Buckingham Palace.

It seemed the whole world wanted an interview with the Flying Family who had endured such a harrowing ordeal and dramatic rescue. Flashbulbs popped constantly. The interviews seemed to have no end. By the time the family boarded the ocean liner *Champlain* (sister ship of the *Isle de France)* to return to America, everyone was thankful for some down time at last.

It didn't take long for the weary travelers to adjust to the contrast between the wilds of Greenland and the luxurious accommodations aboard the huge cruise ship. The Flying Family dined nightly at the Captain's table. After all the media attention they enjoyed the relative calm of shipboard life, and the relief from constantly being in the public eye. Their shipmates were admiring, but respectful, and didn't annoy them (as frequently happened on dry land) by asking for autographs.

Janet and Kathryn enjoyed the rare companionship of other children on the cruise, even though they were teased for being the only kids allowed in the "grown-up" dining room.

But all good things must end, and all too soon the Hutchinsons were deluged with attention from the press once more as they docked in New York City.

Then it was back home to the Governor Clinton Hotel in Manhattan.

"Classroom in the Skies" inside Sikorsky Amphibian
just prior to Greenland crash, August 1932

Janet Lee, Kathryn and Blanche watch their plane sink
September 11, 1932
Life Magazine Cover Photograph of the Year

Janet Lee, Kathryn & Sailor
aboard *The Lord Talbot*
following rescue at sea
1932

Hutchinsons and friend visit
London's Hyde Park

Expedition Crew heads home from Europe
on oceanliner *The Champlain*

15

BROADWAY LIGHTS

The Greenland crash and rescue had made The Flying Hutchinsons a household name. America wanted to hear more of the exploits of these remarkable children and their parents. George wanted to present the adventures of the Flying Family to an avid public while interest still ran high. He also needed to replenish financial stores following the loss of the plane and the mission.

For almost a year, Loew's Theater on Broadway near 42nd Street became their second home. Loew's hosted vaudeville acts and shows much like Radio City Music Hall. The four Hutchinsons became stars of their own stage show dramatizing the crash and rescue. The shows ran for matinee and evening performances, seven days a week. They also had their own nightly radio program on NBC. Life became one big performance.

Two microphones were set up against a huge backdrop of the well-known *Life* photograph of the sinking plane, which filled an entire movie screen behind the Hutchinsons. George and Blanche spoke into one, while the shorter one was shared by the girls. They took turns narrating the suspenseful tale of their crash and rescue to a spellbound audience. Actually, the narrative was pre-recorded and lip-synched by the performers, although it was so smoothly done that no one ever guessed.

The popular show usually played to a full house, sometimes Standing Room Only. Many bows and curtain calls followed the performance. After the curtain was finally drawn, Janet Lee came back out alone, took center stage, and sang "Shuffle Off To Buffalo."

Janet adored the limelight, the applause, the warm waves of audience affection that washed over her with each performance. With her towheaded charm, clear piping voice, and deep dimples, she was a real crowd-pleaser. Once, in the middle of her song, the saxophonist in the live orchestra that accompanied their act got carried away. He interrupted Janet Lee's singing with a long jazz riff. The little girl stopped short, hands on her hips, and glared:

"Hey! Why're you ruining my song? Don't do that!"

The audience roared with laughter.

* * *

Life around a Broadway theater was exciting. The children met famous vaudevillians such as Milton Berle, whose act preceded theirs. Janet's favorite stage act was the Pony Show, naturally, where small Shetland ponies sat on real rocking chairs on a department-store set.

After their last show of the evening, the Flying Family was picked up in a limo and accompanied by a police escort from Loew's to the NBC Studios on Fifth Avenue. The time crunch between their Broadway appearance and the start of their live radio program, *Adventures of the Flying Family*, was that tight.

In the Thirties almost eighty percent of the population owned a radio. Popular personalities such as Jack Benny, Amos and Andy, Fibber McGee and Molly had American families gathering around the trusty Philco with religious regularity. Soap operas dominated the daytime airwaves and heroes such as *The Lone Ranger, The Shadow,* and *Jack Armstrong, All-American Boy* drew faithful followings. America continued her love affair with aviation over the airwaves as well. At one time or another, there were over twenty-five radio serials such as *Smilin' Jack* which featured flying heroes (and one flying heroine).

The Flying Family followed *Jack Armstrong* in a popular time spot. Like most radio serials of the day, their show lasted fifteen minutes. At NBC, the Hutchinsons rubbed elbows with the likes of Jimmy Durante and Rudy Valley. Baby Rose Marie, a popular child singer who later became a regular as Dick's secretary on *The Dick Van Dyke Show* had her dressing room down the hall.

The studio setup was simple by today's standards. The recording room was bare save for a microphone, a big black wall clock, and a sound effects man in the background with his amusing array of primitive props.

There was no lip-synching here, and it was considerably more nerve-racking for the whole family to have to come up with new material and "adventures" to recount each evening to their radio audience.

There was time for only a single run-through. Then the four watched the hands on the big clock sweep inexorably toward airtime, as a disembodied voice intoned: "On the air in ten...in five...in two." Commercials came first.

Then it was showtime, as everyone recited their lines from the script directly into the microphone. George, for whatever reasons, was nervous before these live performances and would ingest almost a whole roll of Tums before airtime. Kathryn was less than enthusiastic about her place in the limelight, as was Blanche, as both were naturally more introverted and private than were Janet Lee and her father. Janet Lee, a natural-born performer, loved it.

As time went on and the details of the trips, the crashes, the adventures had been told and retold, the producers pressured the family to come up with new stories. There was no alternative but to do some inventing.

Janet recalled these improvised stories as "awful" - stiff and artificial-sounding, yet seemingly swallowed whole by a public hungry for new material.

* * *

Thus America's Flying Family lived the life of true media stars in those days, appearing on Broadway seven days a week, brought into America's living rooms on the airwaves each evening at five forty-five. Their smiling faces peered out from boxes of Wheaties, "Breakfast of Champions" and containers of Cocomalt, a hot drink similar to Ovaltine. Loyal fans could send in boxtops from Cocomalt and win a Flying Family jigsaw puzzle in the form of a U.S. Map showing the Hutchinson's Flight of the States.

The family was taken in a limo to meet the Mayor of New York. They appeared in parades. At restaurants, in the hotel, on the streets, the Hutchinsons were approached for autographs. Blanche, though ever a trooper, shrank somewhat from these constant encounters, and Kathryn too found them intrusive. Janet Lee and George were in their element.

Both girls attended Professional Children's School, which was flexible enough to accommodate their schedule. They loved it.

As the months wore on, however, the daily drill became less like fun and more like work. How many times could one repeat the same story, however dramatic, and maintain a high degree of enthusiasm? Eventually of course, public interest waned as well. The family's radio contracts, which lasted thirteen weeks, failed finally to be renewed. The time was approaching for another change.

* * *

While George Hutchinson kept his office on Fifth Avenue and Fifty-seventh Street, near Central Park, he moved the family out of hotels at last and into an apartment in nearby Jackson Heights, and later to a three-story townhouse in Forest Hills. Both Janet and Kathryn began to attend public school, and life began to assume a more normal aspect, on the surface at least. George even directed a school play (Janet and Kathryn had starring roles, naturally.) Blanche became a stay-at-home Mom. Janet and Kathryn sharpened their Morse Code skills by tapping out messages to each other from the 2nd and 3rd floors of the townhouse.

In addition to lecturing and speechmaking, George became an author. *The Flying Family in Greenland*, published in 1935, and *Flying the States*, published in 1937, were used as third-grade readers in the public schools.

During these relatively peaceful years, the girls were often back in Baltimore. Kathryn was especially close to her Grandaddy Hutchinson. Like that talented gentleman, she also loved to paint, write, and play the piano. The girls often stayed with him when their parents were away. Sadly, he died while only in his fifties.

Janet Lee visited frequently with Auntie Nell (actually a cousin) and her husband. Part of the true Batltimore aristocracy, the pair addressed each other always as "Mr. And Mrs. Gillespie." ("Would you like the butter, Mr. Gillespie?" "Yes thank you, Mrs. Gillespie.") Janet remembered her aunt's lemon meringue pies and running down the street to the corner candy store, where for a penny apiece she could stuff her pockets with licorice, jawbreakers and gumballs.

Summer days at their grandmother BB's estate remained the girls' favorite times.

Janet Lee adored BB, and the long summer days spent at the shore with her were full of excitement and delight. She loved to skinny dip with her irrepressible grandmother. Occasionally, BB would take a breast out of her swimming suit, give it a name, and tell a story about it.

As she grew older, BB's eccentricities became more firmly entrenched. She was one of the first women in Baltimore to get a driver's license, even though she always rode around in a chauffeur-driven limousine. One day an unlucky driver managed to shut the limo door on granddaughter Kathryn's finger. An outraged BB hit him over the head with the parasol she always carried rain or shine, fired him on the spot, and left him standing on the sidewalk as she burned rubber driving Kathryn to the nearest hospital herself.

While lots of fun, BB could also be strict with the girls. Etiquette and breeding were paramount, and breaches were taken seriously. A rap on the knuckles with a ruler was apt to be forthcoming if the girls "forgot themselves."

Here, hunkered under the dining room table, Kathryn sneaked a cigarette for the first time. Here too she was caught by her little sister with a beau, necking on the living-room sofa.

When their grandmother didn't much care for a friend or, later, a beau that either of her granddaughters brought home, she had a diabolical way of showing it. BB would have the maid set the table with every single piece of silverware she owned. The profusion of gleaming sterling demitasse spoons, shrimp forks, miniature salt cellars and the like would soon mystify and overwhelm the poor guests. Later, a gleeful BB would point out these deficiencies in etiquette to her offspring.

"Your little friend is obviously just not one of us, dear. Not our sort at all," she'd sniff triumphantly.

* * *

One day out of the blue George received a call from the Head Keeper at the Central Park Zoo.

The Zoo had been Governor the lioness's home for several years following her forced departure from the family's hotel suite at the Governor Clinton. Although Janet never ceased to miss her, she had become reconciled to the idea of her former pet's new life. The family knew that the zookeepers had planned to find a mate for Governor and breed her. Hopefully, their loved pet would live a peaceful life in the zoo, close to her own kind.

Now, it seemed, there was trouble. Big trouble.

The keeper's voice betrayed his deep concern as he explained the situation to George. Governor had, indeed, been mated to a handsome male and in due time gave birth to twin cubs. One, however, was stillborn. Tragically, the remaining cub was accidentally smothered two months later when Governor rolled over on her in her sleep.

"It's the third day we haven't been able to get her to eat," said the keeper in frustration. "No matter what we offer her, she just turns away. That lion's not only griefstricken, she's depressed. If she doesn't start to come out of it soon...well, it doesn't look good."

George glanced down at his daughter's anxious faces; they had come running to crowd around the telephone as soon as they heard it was news from the zoo. Now, they saw the worry in their father's eyes.

"What is it, Daddy?" whispered Janet Lee. "Is something wrong with Governor?"

George held up a hand for quiet and turned back to the receiver. "What can we do to help?"

"I'd say, get yourselves up here as soon as you can. We've tried everything we know. Maybe seeing you folks would cheer the old girl up."

When the girls dragged the whole truth of Governor's predicament out of their father they couldn't get started on their trip to the zoo fast enough. Janet was torn between eager anticipation at seeing her beloved old friend and worry about what they would find.

* * *

But the truth was even worse than they feared. When the four Hutchinsons were finally shown to the cage where Governor lay in a despondent heap in a corner, as lifeless as a fur rug, alarm pierced their hearts.

"Gov! It's us! Don't you recognize us?" "Wake up, girl!" The girls pressed themselves closely against the iron bars.

Slowly, Governor lifted her heavy head. Her golden eyes met their own and she blinked, as if in disbelief. With a sound deep in her furry throat halfway between a grunt and a sob, she lurched to her feet and wobbled toward them. A raspy, pale tongue stretched out and licked the small hands that clutched the bars.

Tears trickled down Blanche's cheeks and swam in George's eyes as they surveyed the condition of their once-vigorous pet. Governor's fur was lusterless and dull. Her ribcage was outlined starkly and the tawny coat sagged on her big frame like an overcoat that no longer fit.

But the little girls were still too thrilled just to be reunited with their beloved mascot for much of that to register. They practically jumped up and down with excitement until the keeper opened the locked cage door with his jangling ring of keys and let the Hutchinsons inside.

With a strangely human keening sound, Governor pushed her great head again and again against the hands that reached out to stroke and scratch her. The lioness had not forgotten her human family.

Janet giggled. "She's trying to wash my skin off!" she said, backing up to avoid Governor's persistent tongue that was licking her cheeks a little too vigorously.

With a satisfied thump Governor rolled over onto her back just as she had when she was a cub, ready for a belly scratch.

For a few moments silence prevailed. There was deep joy in this reunion but sadness and worry as well. How could they possibly help? They couldn't pack Governor up and take her home with them, much as they might wish it.

Suddenly George stopped rubbing the tawny belly and looked up.

"I have an idea."

* * *

After scouring what seemed like every toy store in Manhattan, the Hutchinsons had to admit that toy stuffed lions were in short supply. They picked out a soft fat Teddy Bear instead and bore it back to the zoo.

"Now here's the important part," George instructed, as he proceeded to take off his jacket and rub the fur of the bear vigorously over his skin. "Now you, Blanche," he insisted, passing her the toy. As their mother performed the same ritual Kathryn and Janet got the idea:

"It's so the bear will smell like all of us!" cried Kathryn.

"Me next," insisted Janet, grabbing the toy and rubbing it over every exposed part of her anatomy.

When everyone was finished they re-entered the cage bearing their offering. George held the bear under Governor's nostrils. She sniffed and nosed it eagerly.

"Do you think it'll work?" Janet Lee asked her big sister anxiously.

But at that moment Gov opened her jaws and softly, gently, closed them around the teddy's plump body. She lifted it from George's oustretched hand and promptly turned and trotted with her prize to the far corner of the cage. Turning round and round on the pile of wispy straw, she settled herself down, and carefully deposited her newfound treasure between her outstretched paws. She chuffed contentedly, and gave him a few emphatic licks.

"I think she's accepted it, George," whispered Blanche, as the whole family seemed to exhale in relief.

* * *

Hating to say goodbye again to their old friend, the Flying Family reluctantly made its way back to Maryland.

Soon after their return the zookeeper called once more. Again the family gathered around to hear the news.

"It's hard to believe, but Governor seems like a different animal. She's eating and sleeping again. And she carries that bear everywhere! She licks it and takes care of it just like it's a real live cub. I think we may be out of the woods."

From that point whenever the Hutchinsons managed to visit the zoo, they found Governor in good spirits and still caring for her surrogate baby with a touching devotion. When she finally died at age 13, the teddy bear looked basically just as it had all those years before. Governor had cherished it to the end.

16

HOLLYWOOD

George calculated that the next logical step for America's Flying Family was Hollywood. Movies were fast becoming a national obsession and he believed a wonderful film could be made about their exploits. Accordingly, he began making regular trips to California to try and interest a major studio in the venture. When Janet was nine years old, the whole family joined him there.

Hollywood in the Thirties was a place of power and promise, dreams and heartbreak. The big studios were all-powerful institutions which created the star system. The movie machine churned out film after film to entertain its Depression audiences and help them escape the bleakness of their lives. Moviegoers swooned over matinee idols such as Clark Gable, Bette Davis, Greta Garbo, and Errol Flynn. Elaborate Busby Berkeley musicals and the dance magic of Fred Astaire and Ginger Rogers lit up the screen. Shirley Temple, America's curlyheaded little sweetheart, tap danced and sang to "The Good Ship Lollipop". On the darker side, John Steinbeck's *The Grapes of Wrath* brought the grim realities of the Dust Bowl and the migration of thousands of displaced Americans to stark life on the screen. Desperately, Americans wanted to believe that there were better days ahead.

Onto this stage came the Hutchinsons.

* * *

At first, because Blanche and George still traveled frequently between Los Angeles and New York, the girls attended a boarding school in Azusa, California. At the Maybelle Scott Rancho School for Girls in Orange County the air smelled of orange blossoms and rich earth. In a lush setting of green groves and rolling hills, the sisters enjoyed the sociability of being with other youngsters. Janet jumped right in, delighted with her chance to ride on the school polo team. Her favorite gray pony had one gray eye. Kathryn watered the horses. The girls even liked their new crisp navy blue uniforms. The structure, after so much change, felt comfortable.

Before long, however, problems arose with the boarding school bill. The family was struggling financially since most of George's enterprises were currently still up in the air. Soon everyone ended up back in Hollywood together with the girls in public school.

From their house on Argyle Place just up from Hollywood and Vine the girls could look up and see the huge white "Hollywood" sign sprawled against the dark green hills. This was to be home for three years. The girls loved the little white cottages set around an interior courtyard, and they loved balmy Southern California living. Janet slept, by choice, on the little screened-in porch in the fresh air.

Timothy, a Sealyham Terrier and small but diligent watchdog, was their beloved pet while they lived on Argyle Place. One day, while Blanche had gone out shopping, the girls were sunning themselves in the back yard when Timothy started barking wildly. A strange man was standing in the bushes, exposing himself. The shocked girls ran into the house and locked themselves in. Soon after, someone knocked on the front door. Kathryn stood on one side of the door with a frying pan while Janet stood on the other wielding a baseball bat. It was just a neighbor. Blanche called the police when she got home, but the man was never apprehended.

Life in Hollywood was exciting. Finally Kathryn was able to attend public school as she had longed to do. The California schools were different enough to be interesting, and schoolgirl friendships became important. Janet began attending La Conte Junior High School, while Kathryn entered Hollywood High as a Freshman, rubbing elbows with the likes of Deanna Durbin, Jackie Cooper, and Mickey Rooney.

Every day at lunch period the kids held a sock hop. Every day, a dark-haired Junior girl took the microphone and sang. Soon everyone realized that she really didn't need a mike. Her clear, sweet tones penetrated to the far corners of the auditorium. By the time young Judy Garland became a senior, she had become too famous for public school and was tutored privately on the sets of such movies as *The Wizard of Oz*.

* * *

George tried hard to get a major studio interested in a film version of the Flying Family story. Once it looked as if the deal was going to materialize, but it didn't. Swallowing his disappointment as best he could, George served as aviation consultant on several films.

Most memorably, he worked for a time with the young aviator, magnate and eccentric playboy Howard Hughes. Hughes' passion for aviation and films culminated in his four-million-dollar epic *Hell's Angels.* George found the obsessive millionaire difficult and demanding to work with.

On weekends, the family sunned and swam at Santa Monica Beach. Janet sometimes rode her bike the long twelve miles to the beach. On a Fourth of July outing, she was swimming with a group of children when they noticed a young seal, all by himself, swimming by the boulders at the end of the beach. Janet dove right in after it. Some twenty children followed her on foot along the beach as she pitted her swimming abilities against that of the young seal. Both were fast. Gradually the trailing group of children dwindled down to four. The seal crawled up on the rocks and Janet climbed up after him. He dove, Janet plunged, and the pair came up face to face. Startled, Janet instinctively reached out to hug the sleek little creature.

The frightened seal lunged at her in alarm. Sharp teeth closed on soft flesh. Blood began to trickle down Janet's arm.

"Help! He bit me!" she shouted, floundering through the surf toward the shore leaving a trail of blood in her wake. Two boys grabbed the struggling seal by the tail and hauled it to the beach. Surrounded by the group of children, a bleeding Janet stumbled through the sand up to the highway where they flagged down a passing car.

"My gosh, what happened to you?" The startled driver hustled the little girl into his car and headed for the nearest hospital. The good Samaritan turned out to be a minister, who prayed aloud for Janet the whole way. The pain was sharp, and Janet arrived at the hospital in a state of mild shock. The emergency room was crowded with other children who had sustained injuries and burns from setting off firecrackers.

Alerted, Blanche and George rushed to join their daughter. The doctor insisted Janet needed a tetanus shot, but Blanche refused. She was in her strict Christian Scientist phase and took her daughter home for bed rest and plenty of prayer instead. Janet healed well with no repercussions, and only a slim white scar remained many years later to remind her of that memorable Fourth.

"I just hope those boys didn't hurt him," she mused, over half a century later. She meant, of course, the seal.

* * *

The little family continued to struggle to find its way in Hollywood. Tony Stanford, the stage director who had worked with the Flying Family back in New York, came up with a proposal for Janet. He had always liked her stage presence, and felt that her irrespressible personality would come across dynamically over the airwaves of America.

NBC's *Lux Theater* was radio's most popular drama series in those days. The program featured established Hollywood and stage stars, a full orchestra, and broadcast live from New York for one hour every Monday night. It was a high-class production in every aspect. Whenever a play called for an ingenue, Mr. Stanford began to call on Janet Lee.

Tony proposed that Janet stay with his family when she flew to New York. Rehearsals lasted one week prior to each show. The young actress was paid $500 per performance, plus airfare and expenses.

Janet never actually saw the $500 a week. Still, she jumped into her roles with typical enthusiasm. Over the months she would appear in such productions as *Peter Pan* and *Daddy Longlegs* with well-known actors. Janet co-starred with Gary Cooper, Leslie Howard, George Brent, Anita Louise, Sydney Fox, and child star Freddy Bartholomew.

Janet absolutely loved it. Standing on a box to reach the microphone, her clear, piping tones rang out over the airwaves. She never had an attack of nerves, and stars didn't faze her. Her sparkling personality easily projected into America's living rooms but her dimples, sadly, remained invisible to listeners.

Lux Theater was broadcast from the old NBC Building on Fifth Avenue. One day Janet happened to be sharing an elevator with the young Katherine Hepburn. The actress, upset at some unknown slight, was turning the air blue with some extremely expressive swear words, and Janet stuck her fingers in her ears.

* * *

Janet shared the microphone with Gary Cooper in a production of *The Prince Chap*. She found the tall, gangling actor to be touchingly shy in person. A photograph taken at the broadcast shows Janet perched on Coop's knee while they both studied the script. Actually, the star had trouble with a bit of stage fright, and Janet Lee helped him through it with her friendly chatter. The two shared a huge enthusiasm for horses, and Janet could go on and on about her favorite subject.

The high point of this period came when she co-starred with Freddy Bartholomew in *Peter Pan*. Janet, of course, played Wendy. Both actors were eleven years old. When the brown-eyed young star of such popular films as *Captains Courageous* met his leading lady, she of the sparkling blue eyes and mesmerizing dimples, it was love at first sight.

At first, the two simply stared at each other across the microphone. After Cupid's arrow struck they both made an extra effort to learn their parts by heart so they could drop the script as they spoke their lines and gaze into each others' eyes instead.

Janet's first kiss took place backstage, and she always remembered it.

The morning after the evening performance of *Peter Pan*, Janet Lee Hutchinson was heralded in the trade papers as the "Shirley Temple of Radio." Evidently a lot of heart had gone into this particular performance.

Parting reluctantly after the show, Janet and Freddy schemed for ways to meet again. They corresponded for a time. Freddy, however, lived with an overprotective aunt who had raised him, and said aunt was less than enthusiastic about Janet. The two were never to see each other again.

Freddy Bartholomew was to die at age forty-one, a car salesman. He had been left behind by a Hollywood that easily forgot child stars who had outgrown Neverland.

Janet never forgot her first love. "I never loved anyone more," she said softly, almost seventy years later.

Top:
Gary Cooper and Janet star on NBC's **Lux Theater**

Bottom:
Janet and child star Freddy Bartholomew rehearse *Peter Pan*. Both were 11.

Flying Family
Jigsaw Puzzle

Three Bathing Beauties
Santa Monica Beach, California
June 1936

Cheetah the Chimp, Flash the Collie and the Hutchinsons gather around the Volker, the "Flying Advertisement"

17

GROWING PAINS

As time went on and his Hollywood plans failed to bear fruit, George turned his attention back Eastward. He wanted to get the Flying Family back in the air where it belonged. This time, his vision was global. The ultimate mission for America's Flying Family, as international tensions flared and the thunderheads of war began to build, seemed obvious: the undertaking of a flight around the world to promote world peace. Once more the flying family would bear a scroll for each head of state to sign. As George sought the elusive sponsorship that would be needed for such a huge undertaking, he was forced to leave his family for months at a time. And, in the little house beneath the big Hollywood sign, times were hard and money was tighter than it had ever been.

In George's absence, Blanche Hutchinson struggled hard to maintain her little family. It seemed that hot dogs were on the menu seven nights a week. Luxuries were a thing of the past. But the worst part was that the family that did everything together was together no more. Blanche and George missed each other desperately, and the forced separation took its toll on Blanche.

Janet and Kathryn watched their mother's usually sunny countenance dim as she pined for her husband. In the night they heard her weeping softly into her pillow. When husband and wife were finally reunited, it was hard to tell who was crying harder, Blanche or George.

* * *

Once reunited it was on to Towson, Maryland, where the Hutchinsons settled in an apartment. George continued to pursue his attempts to secure investors in this most ambitious Flying Family project, and to give speeches and lectures.

As Kathryn entered her senior year and Janet began Towson High School as a freshman, life settled into a fairly typical round of school, activities, friendships, and boys.

Janet found public school in Maryland to be disappointingly uninspiring and rigid after what she was used to in Hollywood. "It was like being forced to learn with a knife in your back," she recalled. Still, she managed to excel. By Sophomore year she was elected class president. She was Varsity in track, baseball, basketball, volleyball, and cheerleading. Despite her shortish stature, she even shone on the basketball court. At track meets, Janet piled up win after win in the 50-yard dash. The only day she lost – by a whisker – was the time she slipped next door to her Auntie Kathryn's house before the meet and snuck a glass of wine to calm her nerves.

Kathryn applied herself to her studies, made wonderful grades, and prepared for college. She also became a "boy magnet," which left George less than thrilled. For the first time, he was being forced to confront the fact that his children were venturing into areas beyond his direct control. Even worse, he was no longer front and center in their attentions. Soon both his highly popular daughters were entering the dating arena with a vengeance.

Boys vied to whirl Kathryn around the floor at school sock hops and at Club de la Dance, a ballroom dancing school for teens. Once Janet entered her Sophomore year, she palled around with a group of five girls and five boys. The girls traded off dating each boy, one per weekend, until they settled on the one they liked best.

* * *

Acceptance of change came hard to George Hutchinson. He was strict and rigid with the girls. He grilled their dates as they stammered on the threshold. He glowered through the window as the boys parked in his driveway when bringing the girls home.

Actually, it had all started back in California. George forbade beach parties, but Kathryn snuck off one night and thought she'd gotten away with it, claiming she'd just gone to the movies.

George wasn't having any. "Take off your shoes!" he commanded.

The pile of sand on the carpet gave her away.

Soon it had become a battle of wits, with the girls, especially Kathryn, plotting to outmaneuver their father's machinations. The sisters learned to have their dates park around the corner from their house, because if their father found the pair smooching in the driveway, he'd open the car door and drag them both bodily through the front door. Often Kathryn would look up during a dance at the Valley Country Club and spy her father leaning in the doorway, silently looking on. Then he'd point to his wristwatch and raise his eyebrows. Curfew was midnight, and it was carved in stone.

The girls learned to be ready on time for dates, because if they weren't, George would corner the hapless boy downstairs and grill him mercilessly on every subject from what his father did for a living to his ancestral background to his career plans for the next thirty years.

But perhaps the most extreme, and funny, step George took to protect his offspring was to install a purple light bulb on the front porch. He hoped this would make the girls look pasty-faced and unappealing to their dates, thus discouraging goodnight kisses.

The older the girls got, the more difficulty George had letting go.

* * *

Funloving Janet, always a bit of a rebel, got into some minor trouble in high school. She drank, socially, at parties. One Halloween, a group of kids descended on the two big cannons that banked the local court house and moved them. The police were called, and everybody ran from the scene except Janet. When questioned, she innocently pointed off down the street.

"They went that way!" The others ended up in jail until the next morning, but Janet went home.

Miss Balls, the school Vice Principal, was a white-haired terror and strict disciplinarian to the students. Once, she accosted a defiant Janet with special venom: "You're going to grow up to be nothing but a whore!"

To the teenaged Janet an active social life, including drinking with the gang, was a welcome relief from the kind of driven, intensely high achievement goals she set for herself.

At this point her drinking was social, but sometimes even then she couldn't predict what might happen when she drank.   One morning she woke up lying in the gutter in front of a friends house following a wild party the night before. She had no clue how she'd gotten there.

Janet had had only one Zombie to drink at her graduation dance when she dove off the high diving board at the country club in her formal. The ladylike Blanche, beholding her abashed daughter in soggy crinolines, had trouble believing her eyes.

* * *

George Hutchinson still spent a great deal of time away from home during these teen years, and while he was gone the three females got along harmoniously, for the most part. When George returned, however, everything changed instantly. The house of women became a patriarchy once more.

For Janet and Kathryn, it was difficult to adapt to George's stricter rules and sometimes over-the-top control tactics. It was also hard to switch from being the center and focus of their mother's attention to the more peripheral place they occupied in George's presence.

Blanche and George, always physically affectionate and demonstrative with each other, cuddled and laughed and kissed. They drew into their romantic universe of two.

All teenagers consider themselves the center of the universe, and in this household that space was occupied by George. He liked to micro-manage the household, as well. Cleaning, organizing, cooking, daily routines were no longer good enough. Everything had to be reconfigured to please the Lord of the Manor. Blanche almost never directly confronted her husband when she disagreed with him. However, she employed the "hand in the velvet glove" technique and often eventually got her own way.

George Hutchinson was the kind of man who both exuded and consumed a great deal of energy. The atmosphere in a room shifted, perceptibly, when he entered it. He courted and expected attention from those around him, and when that didn't work, he demanded it.

The old relationship Janet and Kathryn had with their father, a combination of love, trust, hero-worship and a firm belief in his omnipotence, could not survive their adolescent years intact. Love locked horns with a chafing resentment as their drive for independence vied with George's need to control. From this point they would continue to love and respect their father, and at least appear to defer to him. But Janet and Kathryn were both strong personalities in their own right. Life in the Hutchinson clan could no longer reflect the idyllic All-American picture George wished to paint.

* * *

When the family was living in a Baltimore apartment complex, a commotion was caused one night by a fire in the buildings' parking garage, underneath the main structure. The fire was especially dangerous because the automobiles parked in front of the raging flames could combust, their gasoline engines exploding. Firemen would not venture into the garage to move the vehicles, deeming it too dangerous. But George Hutchinson eluded the barriers, charged into the garage, and singlehandedly pushed several vehicles out of the garage and out of the fire's path.

Blanche was beside herself with worry, but she knew that her husband, as always, would not be deterred. Spurred on by the older man's example, the firemen ran into the garage and were able to extinguish the flames, which had begun in the building's utility area adjacent to the garage.

* * *

Janet's ongoing love affair with horses was a dependable source of escape and solace during these difficult years. Always a lover of all things equestrian, Janet had learned to ride at age five. Thanks to BB, the little girl had been enrolled at a Maryland riding academy.

The lessons progressed far too slowly to suit her. She insisted "I'll teach myself," and proceeded to do just that. She was given the tiny pony Tweedledee to ride. Tweedledee had thrown her repeatedly for a week, but by Week Two she was putting her mount through its paces like a pro.

When she was sixteen and a senior, Janet was at last given her own horse for Christmas by her beloved Auntie Kathryn. She named him "Laudy", short for Fort Lauderdale, the Florida city where her Auntie lived. She boarded Laudy out at various farm stables.

Janet always cherished the book she created for a school assignment in 1938, titled "Horses: Their Story." Dedicated to Will Rogers, who "loved horses as much as I do," the book covered the subject of the horse exhaustively, from prehistory through the present. Janet included fascinating and little-known tidbits about the history of the horse, the care of the foaling mare, riding techniques, etc. She concluded her treatise with "A Merry Chase", a story she wrote herself. The story described her first fox hunt, a day she described as "the happiest in my life". The experience made her feel "as though I owned the whole world."

* * *

As often as she could, Janet rode out alone after school to an isolated strip of land near the horse stable called Sandy Bottom. Sandy Bottom was a stretch of open, grassy land down in a little valley. It was a place where Janet could give Laudy his head and let him run, a place where she could be alone with just the wind and the sky and her beloved horse. It was sheer heaven.

But even paradise has its serpents. One day two men in derby hats materialized over the top of the ridge. They ran directly at the startled girl, lunged for and and grabbed Laudy's reins.

"It was like that scene out of *Gone With the Wind*," Janet remembered. "Where Scarlett gets attacked. Laudy reared straight up. Withoung thinking I slashed down at them with my crop, hard as I could. Laudy hit one of them with his hoof as he reared up. They let go."

Laudy took off at a wild gallop. Janet was shaking as she unsaddled her horse that night and went through her regular chores at the stable. She knew what a close call that had been. She could've been raped, killed or both.

Shaken or not, she never told. The fear of being deterred from riding was greater than the risk.

"I don't remember feeling fear," Janet mused. "Just anger. The anger made me strong."

Janet became very angry with her father once and stormed out of the house. When she didn't return by dusk a worried family started searching. She was found at last at Laudy's stable, curled up by the horse's side on the hay and sleeping soundly.

* * *

While his daughters were living the lives of fairly typical American teenagers, George Hutchinson was working to make his biggest dream yet a reality.

George had persistently sought the funding he needed for the flight for world peace, his most idealistic and ambitious project. He was turned down time and time again (87 times, to be exact) and still refused to give up. Finally, in Pittsburg, he found funding and support from a wealthy steel magnate known to the girls as Mr. Iverson. An excited Flying Family flew to Pennsylvania to meet with their new patron.

The Iverson's home in the suburbs of Pittsburg could only be described as palatial. As they were warmly welcomed by their host and given a tour of the premises, Janet was especially taken with the magnificent dining room table, long enough to seat an army, set off by a cascading fountain in the middle. Janet's love of fountains persisted from that day forward, and years later a myriad of fountains filled her own Ponte Vedra Beach home with the gentle, trickling sounds of water on stone.

The big plans continued to gel. As before, signatures of each head of state would be affixed to a scroll. This huge, intricate document bore the likenesses of the ruler of each country and was painstakingly prepared under a microscope. It cost upwards of $10,000 to create, and was insured for one million dollars.

Preparations for this All Nation Flight, naturally, were expensive and time-consuming. Kathyrn and Janet became New York City natives once again as the family journeyed back to the city. Both young women attended Morse Radio School, Janet, at age fourteen, becoming the youngest person ever to earn a Radio Operator's License. School was located in the RCA building, where, just upstairs, an upstart invention called *television* was being developed.

* * *

The sojourn in New York was exciting but full of hard work. Given more freedom, the girls learned to use the subway to commute from radio to language school daily. At the Berlitz School the girls learned to recite the message of world peace in French, German, Italian, Spanish, and of course, English. Because it was their job to carry this message to each country visited, the sisters remembered the words their whole lives:

We the leaders of the nations of the world proclaim our greetings and good wishes to all peoples in the fervent hope that this document may serve to strengthen world fellowship for the greater glory and welfare of humankind.

* * *

The plane, a Lockheed Electra 10A, held a crew of twelve and was well appointed and comfortable. Once again, Pete Redpath served as navigator. Lee Dyce was the copilot. The girls' Morse radio skills would be employed on the trip as needed.

Franklin Delano Roosevelt became the first world leader to sign the scroll in April of 1939. After a big sendoff from Washington, the Flying Family flew directly north to Ottawa, Ontario. Then it was south, to Cuba and the Caribbean, and on to Mexico and South America.

Once again, Janet's young life became a whirlwind of new places, people, and experiences. Spending an average of three or four days in each country, the girls grew used to a routine of waking up each morning, trying to recall what country they were visiting, and charging forth into a day crammed with new adventures. On an average day the sisters might be squired around capital cities on sightseeing tours and presented to heads of state. They posed for endless photographs. The two attractive, vibrant young women were a magnet for the press as they posed on the beach at Acapulco or in formal dress at parties. In the evenings, balls, dinners, and social events galore left their heads whirling.

Strangely enough, many of the South American leaders just happened to have eligible and good-looking sons. These young men vied for the chance to dance with the pretty young Americans. Later, Janet and Kathryn would accuse each other of having the "burnies" for one young man or another.

* * *

As the trip progressed the countries tended to blur one into another, but certain vivid moments stood out.

To reach Le Paz, Bolivia, the family flew over the second highest mountain in the Andes. The airport was located on the edge of a cliff, and when the heavily-loaded plane took off again it dropped like a stone down into the valley, leaving everyone's stomachs where their throats should be. The native population, the girls noted, all had big chests because extra lung capacity was needed to breathe the thin air. The North Americans were left gasping for breath.

The primitive and impoverished conditions made a deep impression on the girls. Women in derby hats, smoking cigars and riding donkeys, filled the unpaved streets. Janet reached out to pet a llama, and was spit at for her trouble.

"Rojo," or Redhead, the natives whispered, reaching out in superstitious wonder to touch Kathryn's flaming auburn hair.

Flying over Ignasu Falls, Brazil, the world's second highest waterfall, Janet and her father both photographed the stunning sight. Later, Janet was delighted when she discovered that the photos she took with her sturdy little Brownie were better than her father's, taken with an expensive camera.

In Bogota, Columbia the Hutchinsons visited an exquisite cathedral decorated with gold and marble. Looking down, they saw they were standing on a bare dirt floor. There too the girls attended a dance at a military academy, where Kathryn had her first social drink (Scotch) and promptly decided she hated the perfume-like taste.

The girls enjoyed observing the Sunday courting rituals of the local Columbian youths from the veranda of their hotel. Dignified chaperones accompanied each young woman as they strolled slowly around the square in front of the hotel while a band played from the gazebo. The local swains lingered on the fringes. When they saw a girl they wanted to meet, they politely asked permission of her chaperone. The couple would then parade solemnly, in tandem, around and around the square.

In the capitol of Paraguay the family awoke to find itself in the middle of an attempted military coup. Soldiers poured through the streets outside their hotel amid shouts and gunshots. The Hutchinsons were only too glad to be back in the air and away.

In Caracas, Venezuela the family was ushered into a suite overflowing with orchids (which grew like dandelions there) and exotic fruit. The sisters felt like pampered princesses among the lush beauty.

* * *

An earthquake in San Jose, Costa Rica supplied the scariest moments of the trip. Janet and Kathryn had gone to the movies to see Bette Davis in *Dark Victory*. Right in the middle of the tearjerker, a tremor struck the little theater. Startled moviegoers shook in their seats as the very foundations beneath their feet shifted. Everyone ran for the exit, except for Janet, who was still engrossed in the movie and had to be pulled to her feet by her sister.

The girls ran back to their hotel to join their parents, but the hotel was evacuating and they were immediately hustled outside as still more tremors struck. George, caught in the bathroom, was the last guest out.

Horrified, the family watched buildings sway and masonry topple. A huge fissure opened before them in the earth and the girls thought they saw people slip into the gaping hole and disappear. Frantic, everyone turned and ran back into the hotel for shelter. Blanche hustled the girls under the frame of a doorway for protection and began praying the Lord's Prayer out loud. Slowly, the quakes subsided. Damage to the hotel was so extensive that later it had to be torn down.

Once more it was good to board their plane and leave chaos behind. Yet, just as the Western Hemisphere leg of their trip was completed, a more dire event put the hiatus to the Flying Family's last trip and most ambitious venture.

Hutchinsons with "Scroll Of All Nations"

Flying Family prepare for takeoff on All Nations Flight
1939

Janet & Kathryn May 1939
Cocacabana Beach, Rio
All Nations Flight

"Rosie the Riveter"
Janet modeling, 1943
Senior in High School

18

THE WASP YEARS: THE CALL

There was not much that could stop George Hutchinson in his tracks, but World War II managed to do the trick.

The All Nations Flight was cut short as global war broke out and a sorely disappointed Flying Family returned home. World peace became a dream that must be laid by, although it would continue to be a driving passion throughout George's lifetime.

Back in Maryland, the girls resumed their schooling. Janet went back to high school, Kathryn on to college. Janet's junior and senior years were full of friends, dating, sports, and normal teen activities. But already she was looking toward the future.

It was clear to the energetic young woman, even then, that an ordinary life was not for her. Once she saw the newspaper announcement of the formation of the Women's Airforce Service Pilots (or WASP), her gaze was set in one direction: skyward.

* * *

The decade of the Forties, when Janet would reach young womanhood, were years dominated by World War II. In the United States, although war production was pulling the country out of the Great Depression at long last, the average annual salary still hovered under $1300. Only recently had the Supreme Court ruled that blacks had the right to vote. Life expectancy for a female was only 68 years, 60 for males. Only about 55 % of American homes had indoor plumbing.

The young people of the nation danced to Tommy Dorsey and Glenn Miller's Big Band sound or swooned to Frank Sinatra or Bing Crosby. Jitterbug, forerunner of Rock, got the kids up and jumping to the music. Radio was the lifeline for Americans in this decade, connecting them to the greater world for vital wartime information. The war boosted the popularity of motion pictures as never before. The War Office itself pronounced movies essential for morale-building and propaganda.

In Hollywood all heroes were handsome and strong and their heroines were all beautiful and worth fighting for. Germans and Japanese were cast as one-dimensional villains. Films such as *Casablanca, Mrs. Miniver, Lifeboat, The Best Years of Our Lives* told the emotional tales of Americans affected by global war.

The image of the American woman became more complex, changing to meet the demands of the time. On the one hand she was the Girl He Left Behind, the wife or girlfriend who kept the home fires burning. Feeding the national penchant for fantasy, images of delectable pinup queens such as Betty Grable or Rita Hayworth graced thousands of locker doors.

But wartime meant exposure to the bigger world and more freedom for American women, whether in the military as WACs or WAVEs or on the production lines to help fill wartime quotas. When "our boys" shipped out, women stepped up to the plate. By 1943, virtually all *single* women were employed. (Only then were married women allowed to work.) Rosie the Riveter became a national icon.

Change was in the wind.

* * *

We live in the wind and the sand...
    And our eyes are on the stars.
        WASP Yearbook, Class of 1944

There was a time, during the years of WWII, when courageous women blazed the trail into the skies for their sisters to follow. Their story, the "best-kept secret" of WWII, is that of the WASP or Women's Airforce Service Pilots.

The saga of the WASP began as a result of the severe shortage of pilots during the early years of WWII, as increasing numbers of male air cadets were drafted into combat duty. Qualified pilots were needed to fly all non-combat missions in support of the war effort. Females were called on to ferry personnel, haul cargo, deliver aircraft to points of embarkation, test fly new, old and rebuilt planes, and tow targets for gunnery practice (dodging live ammunition.)

Two prominent female aviatrices, Jacqueline Cochran and Nancy Harkness Love, foresaw the need for women to step into the gap. The months that followed the creation of the WASP proved that women could accomplish things on the ground and in the air that few had deemed them capable of.

Before the WASP venture was over, a select and superbly trained coterie of women would fly every type of Air Force craft - trainers, bombers, and fighters. The WASP would test fly planes in questionable condition and in situations that some male pilots had refused as too dangerous. Thirty-eight WASP would give their lives in the service of their country.

Unfortunately, most Americans remained unaware of the WASP effort throughout the war. The encompassing drama of a world at war, and concern for the plight of "our boys," left little attention to spare. After the women's job was done, a conspiracy of silence prevailed. To this day, most of the world remains unaware that the WASP ever existed.

* * *

It took Janet Lee Hutchinson less than one minute from the time she read the newspaper article announcing the formation of the WASP program to full-fledged commitment. This was something she simply had to be a part of. The year that followed became the happiest time of her life.

The WASP adventure provided a chance for the 18-year-old, still overshadowed by the figure of her dominant father, to prove her mettle outside the auspices of the Flying Family. Here Janet would for the first time meet other young women who were as adventure-loving, fearless, and in love with flying as she was. Here she would form friendships that would last a lifetime, and learn to grieve for the loss of a dear comrade. She would find herself locked up in a roach-infested West Texas jail, facing the worst humiliation of her young life. She would be driven to the depths of despair and to an action so shame-ridden she could never share it with the family with whom she was so close. She would learn that goals met and success attained, symbolized by those coveted and dearly won Silver Wings, would not insulate her from pain. She would leave girlhood behind and become a woman.

* * *

With singleminded determination Janet set out on her course. Immediately following graduation she signed on with the Signal Corps at Bendix Radio, inspecting radio equipment for the government. With her salary she enrolled as the only female flight student at the Westminster Airport north of Baltimore. She would need eight hours of training to enable her to solo, a prerequisite for the WASP.

On a late winter's day in 1943, Janet sat at the dining room table, gnawing on the end of a pen. The application for the WASP program was spread out in front of her. She shifted impatiently in her chair. She'd passed the grueling physical, just squeaking in under the height and weight requirements. Now she was intent on finishing the written form. She *had* to be accepted. Any other outcome was unthinkable.

"Almost finished?" George approached from the living room where the radio was broadcasting one of Roosevelt's fireside chats, and laid a hand on her shoulder.

Janet glanced up. "Yep. Going to get this thing in the mail first thing tomorrow."

George pulled out a chair and sat across from his daughter, smiling knowingly. "You're a shoo-in. You know Jackie Cochran and I did some fancy flying together in England. Did I ever tell you about that bombing raid in London when-"

Janet grinned. "When you were stuck in the bathtub, you mean?"

"It was incredible. The air was full of screaming sirens, and the lights went out in the hotel, and before I knew what was happening - BOOM! A big one hit us, and that old iron tub turned right over on me."

"But, true to the Hutchinson luck, you escaped without a scratch, right?"

George's eyes twinkled. "And just when did you get in the habit of finishing my stories for me?"

* * *

From its conception, the WASP was a result of female vision. Two remarkable and very different pioneer aviatrices shared in its founding. Nancy Harkness Love, who had founded the WAFS or Women's Auxiliary Ferrying Squadron as a temporary, civilian women's flying corps in 1942, and Jacqueline Cochran, creator of the Women's Flying Training Detachment classes, were thrust into a position of mandatory cooperation.

In 1943, the Women's Airforce Service Pilots, or WASP, was formed as these two organizations were joined under the direction of General Hap Arnold into a united force, with Cochran as director.

Nancy Harkness Love began to fly in 1930 at age 16, later operating a successful aircraft sales business with her husband Robert Love and going on to fly planes to Canada for shipment to France. Recognizing the need for pilots to ferry aircraft as World War II approached, she handpicked highly experienced women pilots and formed the WAFS.

Meanwhile, Jacqueline Cochran, America's foremost woman pilot (eclipsed in fame only by Amelia Earhart) was working to create something extraordinary. Her vision was to bring together a corps of seasoned women pilots and train them to fly "the Army Way," replacing male pilots being sent overseas. In many ways the story of the WASP is the story of Jackie Cochran.

* * *

Jackie Cochran's rise to fame was a true American fairytale, with one important difference. This heroine did not retire with the Prince to live "happily ever after" but went on to carve out a life of remarkable achievement.

Jacqueline had started life as a foundling. Adopted into a poor family, she grew up working in the sawmill camps of North Florida and South Georgia. From these inauspicious beginnings she struggled to constantly reinvent herself as she went through life.

Starting out as a hairdresser in Montgomery, Alabama, the attractive blonde was soon working at Antoine's in Miami Beach. Here she met tycoon and Roosevelt administration New-Dealer Floyd Bostwick Odlum. Odlum, a self-made millionaire, soon became embroiled in a scandalous divorce. In the settlement his ex-wife ended up with New York's Bonwit Teller department store.

Ignoring popular opinion, Odlum became Jackie's greatest supporter, encouraging her to learn to fly and assisting her in the startup of Jacqueline Cochran Cosmetics. In 1936 the two were married. By 1939 Jackie Cochran had become a national sensation. She had her own successful business, a powerful husband, an enviable place in high society, and several flying records. The rags-to-riches story of her origins had become an asset in the Cochran myth.

Jacqueline earned national attention in 1937 when she took first place in the women's division of the Bendix Air Race, breaking Howard Hughes' New York to Miami speed record in the proccess. The following year she set a new woman's coast-to-coast record. She had only had her pilot's license for five years at the time. But Jackie's vision ranged wider and higher.

* * *

In 1939, war loomed as Nazi tanks invaded Warsaw. Probably it took someone with Jackie's determination, the grit to rise from backwoods sawmill camps and grinding poverty to a life of accomplishment and wealth, to even begin to fight the fight she now faced. She intended to convince the male military establishment that they needed a special corps of women pilots who could deliver the goods every bit as well as men.

She faced an uphill battle. In September 1939 Jackie sent a letter to First Lady Eleanor Roosevelt, outlining the case for the WASP. The forthright Mrs. Roosevelt immediately threw her support Cochran's way in her newspaper column.

"This is not a time when women should be patient. We are in a war and we need to fight it with all our ability and every weapon possible. Women pilots, in this particular case, are a weapon waiting to be used."

Jacqueline met and attempted to win the support of General "Hap" Arnold, Commander of the Army Air Forces. General Arnold put her off at first and recommended more training and preparation. Cochran ended up flying bombers to Great Britain and assisting the Air Transport Command (ATA) by ferrying planes during the aerial bombardment of Britain. It was at this time that she met (Honorary)

Colonel George Hutchinson, who was also serving his country as a volunteer by flying war planes to Europe. (George, having been granted the honorary rank of Colonel by the Governor of Texas following the flight to the 48 states, used it religiously from that point on.)

It took Pearl Harbor and a slowly-recognized crisis of manpower for the powers-that-be to finally realize that the WASP was an idea whose time had come. Eventually Cochran's persistence won out and General Arnold threw his support behind the WASP with wholehearted commitment.

Rivalry, misunderstandings and complications remained concerning leadership of the new corps. Nancy Love's vision for her elite squad of highly qualified ferrying pilots was more limited than Cochran's sweeping goal of training a large number of female pilots to fulfill all functions. In the end, Cochran prevailed, with Love maintaining directorship of the ferrying squad but serving under Cochran. In 1943 the two distinct organizations, Cochran's Women's Flying Training Detachment classes and Love's original WAFS were combined into the WASP. The seeds for the first all-woman flight training base was born.

Cochran herself would be emphatically in control, securing for herself the directorship of all women pilots both in training and on duty. She believed that a woman-directed female corps would be less likely to get lost in the frequently hidebound male military hierarchy. Jackie wanted to have all her bases covered. From the beginning she ran a tight ship and put herself *in loco parentis* to the young women under her command. "Her girls" would be carefully handpicked, checked and supervised. She saw the need to reassure the American public that these young women would not be running wild on Army bases. Discipline would be strictly enforced.

* * *

Young American women heard the call, and responded with unprecedented enthusiasm. Some were seasoned pilots already flying commercially . These women would have the worst washout rate, struggling to mentally override flying techniques that had become routine to them and learn to fly "the Army way" instead. Many, like Janet, had paid for flight instruction in order to meet the minimum 35 hours training time they needed in order to qualify for the program.

Some 25,000 applications poured in by mid-1944. Of these, 1,850 were eventually selected. The washout rate in training for these enthusiastic young women would be close to fifty percent.

Cochran's WFTD had been formed in November of 1942 and began training the first group of women at Howard Hughes Field in Houston. The facilities there were makeshift, outdated, and inconvenient. Women had to hike half a mile to a bathroom. The trainees practiced in old, dilapidated planes. With a call from the U.S. Air Transport Command for upwards of 1500 women pilots before the end of 1944, new arrangements for a training base had to be made, and fast.

Avenger Field, outside Sweetwater Texas, was selected to be transformed from a base for male Air Force Cadets into the only military flying school for women in the U.S. Situated smack in the middle of the empty spaces and skies that go on forever in West Texas, Sweetwater, a small town seventy miles north of San Angelo and forty miles west of Abilene, was given its identity by the proximity of Avenger Field.

Shaped roughly like the state of Texas, Avenger Field consisted of long, low, drab-gray buildings and narrow taxiways. Surrounded by flat plains, mesquite, and buffalo grass, Avenger provided a wonderful setting for airplanes but a bit of a challenge for humans.

The near-constant winds whipped up storms that could shut down visibility with a solid-appearing red wall of dust. Sweltering summer temperatures could reach as high as 100 degrees Farenheit by April, and stay that way through September. Winters could be wet and penetratingly cold. Yet by 1944, hundreds of eager young American women made their way to Avenger, new home of the WASP.

19

THE WASP YEARS: TRAINING

When Janet was inducted into WASP Class 44-6 in January of 1944, she was, at age eighteen and a half, the youngest in her class. She was also the smallest, squeaking in under the height standard at 5 feet, 3 ½ inches tall. Her flight instruction hours barely met the 35-hour requirement, with her log book recording exactly 35 hours plus 20 minutes instruction time. But she had been accepted, and she was chomping at the bit.

Except for extended stints at summer camp, Janet had never lived away from home before. Yet it was the pure adrenaline of excitement, not fear, that coursed through her veins during the long train ride from Maryland to Texas. The world seemed big, shining, and new, and she was more than ready.

Janet was almost too excited to eat, but she forced herself to sit down to a good breakfast in the dining car. The server set a plate loaded with food down in front of her. Eggs, bacon, toast, fine. But what was this lumpy white mass, oozing grease, that had been plopped down next to her eggs?

"Uh, Ma'am?" Janet spoke up. "I didn't order this. What is it anyway?" she asked, pointing at the questionable mound.

The serving girl followed her pointing finger, then laughed.

"Honey, that's grits. And you don't order grits. Grits just comes."

Janet never did become a fan of grits.

After disembarking from the train in Austin, a crowded bus collected Janet and the other fledgling WASP for the remaining trek to Sweetwater. As Janet found her seat, the sound of excited young voices rose around her.

"I hear it's pretty basic, the setup and all. At least we'll have bathrooms - not like those poor girls in Houston, that had to hike a mile to the showers."

"I just hope to God my bunkmates don't snore."

"I hope I didn't bring too many suitcases!"

Finally the bus lurched to a stop and Janet climbed down the stairs of the bus. She stood taking in her first view of Texas. The plains stretched out before her, wide open spaces as far as the eye could see. A robin's-egg-blue chilly January sky seemed higher, and vaster, than skies back East. She took a deep, slow breath and inhaled sage and dust. In the distance she could see corrals and horses grazing. *Cowboy country*! Janet loved it from that moment on.

Janet liked Hally Stires, her Staff Adviser, from the first moment they met. Hally greeted the new arrivals with a big smile and a tour of the facilities, starting with the Administration Building.

Janet squinted up at the strange, 10-foot wooden figure perched atop the building.

"What in the world is that?"

"That's Fifinella," Hallie replied.

"Fifi-who?"

Fifinella, a cute and curvy female gremlin, was a mascot created especially for the WASP by the folks at Walt Disney. Her purpose was to "scare away" the male gremlins widely thought to be the mischevious cause of plane malfunctions. Sexily clad in gold tights and red boots, the cutout figure added a bit of whimsy and perhaps made the new environment appear less intimidating. Her emblem appeared on flight jacket insignia and on stationary. Later, WASP alumnae would become part of the "Order of Fifinella."

Avenger was rather overwhelming in its vastness, as Janet described in her first letter home to her parents:

There are more girls and planes than I've ever seen in my life...this place has everything, hangars, mess hall, barracks, a huge gym and outdoor athletic field, theater, post office, administration building, hopsital and a wishing well to throw money in when we have a flight check.

She would find out that the well also served as the scene of an "initiation by dunking" ritual following each trainee's first solo.

Janet's first glimpse of the Stearman PT or Primary Trainer the recruits would be flying was equally daunting.

"It looks huge!" Janet whispered to the girl beside her. All of a sudden, she felt like a little girl again, measuring herself against the big inflated wheels of her father's Sikorsky. *How will I ever fly that thing?* she thought.

* * *

WASP trainees included a ballet dancer, a member of Radio City's Rockettes, and a movie star. The bright young women who thronged into Avenger came from all walks of life. There were professional athletes, like golfer Helen Dettweiler, as well as debutantes and heiresses, such as Marion Florsheim of the shoe fortune. There were students, models, and doctors. And of course there were aviatrices such as Helen Richey, the first woman pilot to fly for an airline, and Janet Lee Hutchinson of Flying Hutchinson fame.

The women were married and single, with and without children. A few had lost husbands in combat, or had prisoner-of-war spouses. Some were already flight instructors. One, Hazel Ah Ying, was Chinese. (Ah Ying joined the WASP after being turned down by Chiang Kai-shek's airforce in the Chinese struggle against the Japanese.)

All, like Janet, loved to fly. Few were timid or retiring. Yet, in a day when the term "feminism" had yet to be coined, they didn't see themselves as pioneers on the forefront of Women's Liberation but simply as women who wanted to fly airplanes, and serve their country in the bargain.

For many it would be their first time away from home. Paying their own way, they came by bus or car, hitchiked, rode the train or even flew. They arrived with fur coats and poodles, or cardboard suitcases. For the cross-section of young women who gathered here from towns and cities across Ameica, the dust-deviled and windswept plains around Avenger would become the stage for what most would claim to be the most exciting, challenging, arduous, and single best year of their lives.

Even in flight school, Janet had encountered no other women much like herself. Here, she felt herself instantly among her peers. Lifelong friendships would be forged at Avenger, based on the mutual passion of flight combined with the closeness that proximity to disaster and death can bring.

* * *

The women bunked in eight barracks divided into two bays. Six baymates shared cots in a single room. Janet's baymates included Patricia Hopkins, Dorothy Hines, Jean Hixson, Mary Hillberg and Margaret Hicks - all fellow "H" surnames. Intimacy developed quickly in the face of the close living arrangements.

"Holy smokes," Janet breathed as the six dumped their heavy luggage beside their narrow beds. Much of what they had brought would end up being stored or returned home for lack of space. It didn't take long to take in the bare-bones basic layout. Cots were so narrow that the girls discovered they'd better turn over carefully to avoid a nocturnal spill. At the foot of each bed was a footlocker, serving as closet, dressing table and bureau. The one bathroom was shared with the adjoining bay and contained two stalls, two sinks, two showers and one mirror.

Patricia ("Patri" to her friends) would become one of Janet's closest buddies, as would Dorothy or "Dottiebelle." Patri rolled her eyes as she surveyed the single mirror and the naked lightbulb hanging from the ceiling.

"And for this they're docking our pay a dollar sixty-five a day?" she sniffed. Trainees earned $150 a month, less $1.65 a day for room and board.

Dottiebelle was vivacious, blonde, and funloving. "I guess they don't want us to spend a lot of time putting on makeup in the mornings," she said regretfully. She was known for being something of a glamour-girl.

"If we can even *find* the mirror at six o'clock in the morning," Janet grumbled.

The baymates that would become Janet's family for the next seven months shared everything. Clothes, makeup, gossip, letters from home, nothing was off limits. When she wrote to her folks,

Janet's letters were full of praise and affection for the "swell gals" with whom she was becoming so close. One of her buddies, a bit of a Wild Child, had to be rescued from compromising situations on a regular basis. One night she stripped down to the buff on the athletic field and had to be unceremoniously grabbed and hauled home. It was all for one and one for all.

Janet, as the youngest and most petite, became something of a mascot to the outfit. Only much later did she discover that Dottiebell had been a bit put off by her initially.

"Actually, I thought you were a spoiled brat - always the center of attention!" she laughed. "But you're OK, Hutch." As in any secret society, nicknames and in-jokes abounded.

Regardless, Dottiebell and Patri ended up mothering Janet, and eventually were to pull their bunkmate through the darkest days of her young life.

* * *

Everywhere the trainees went, they marched, usually to a song. These self-created songs and chants built morale, even though Jacqueline Cochran sometimes bemoaned their raucous lyrics.

Oh, I'm far from home
Where the wild Texans roam
Where the snakes and tarantulas play
Where seldom is heard
An encouraging word
And we never have time to make hay.
A WASP trainee am I--
All sunburned and dusty and dry.
There's no time to play,
They work us all day,
Volunteers, but we'll never know why!

If I graduate
I'll get out of this state
And never see Texas no more--
We'll ferry their planes
Through the wind and the rains
And help all our boys win the war.

The new recruits learned fast that any emotional support available would be provided by each other. Their training program was the same as the one male cadets were put through all over the country. The day began with reveille at 6 A.M., a scramble out to the field for daily calisthenics, strict room checks that included making up their beds so tightly that dimes actually bounced, breakfast in the mess hall. Then, the work began. Ground school in the morning, followed after lunch by flight school, cross-country flying, instrument flying, and training on the link flyers. After dinner the girls worked on drawing up flight plans, homework, or night flying.

Trainees determined to stay the course quickly learned to adapt to the harsh but exciting life at Avenger. Phenomenal summer temperatures and a near-constant wind that blew sandy grit into their hair, skin, and clothes did not dissuade them from sunbathing at "Avenger Beach," a sun-scorched patch of dirt between the barracks. These were the days before concern about skin damage, so basting oneself in the merciless Texas sun was a favorite way to unwind. One trainee proudly boasted in a letter home that she was getting thinner, fitter and tanner than she had ever been.

Frequently the nights were so hot the girls opted to move their cots outside and sleep under the high Texas stars. Snakes, however, made this a risky proposition. The occasional rattlesnake was found curled up in the undercarriage of the planes.

Rains could become so heavy in the spring and winter months that planes were grounded. Cold penetrated, especially at night. Sheepskin-lined leather jackets protected them from the chill in the cockpit.

Extreme heat was harder to avoid and of course air conditioners were unheard of. Modesty often went by the board in the close quarters and the heat. Somewhere a photo exists of a stark-naked Janet sweeping out the bay for inspection.

* * *

Clothes did not make the woman at Avenge Field. Initially the WASP had no official uniform. Instead, the women were issued Army surplus mechanic's overalls, size 44 and up. Draping themselves in the folds of what came to be known as "zoot suits," cinching in waists and rolling up sleeves, the women learned to joke about it. The zoot suits, along with leather flying gear for cold open cockpit flying, flight caps and goggles, comprised the women's regulation flight gear. For more formal occasions, early classes dressed in white short-sleeved shirts, men's khaki trousers altered to fit WASP anatomy, and a general's cap.

The women joked that they had to spend more time changing their clothes for each activity than actually flying. Only in 1944 did Jackie Cochran, after a year of negotiations, procure the official WASP uniforms of Santiago Blue that would so strikingly set off those coveted silver wings. The hat was a modified beret (which Janet detested).

Janet wrote home soon after arrival to ask her mother to send her Dad's old flight boots, as they were all out of boots at Avenger. It would also save money.

The day the package arrived, her baymates gathered around to watch her try them on. Janet wiggled her toes freely inside the roomy leather.

"Looks like there's plenty of room in there, Hutch," Dottiebell said.

"Big bootsteps to follow in, right, kiddo?" added Patri. Perhaps they already sensed that Janet would knock herself out trying, as always, to fill her father's larger-than-life shoes.

All in all, it was a life of exhilarating challenge. The women got no coddling. To most complaints they were told "That's tough," in true Army tradition. Although the women were of Civil Service rather than military status, they were treated like soldiers.

* * *

The flying itself made all the exhausting hours, all the heat and cold, all the putting up with harsh instructors, worth it and more. The women sang,

> *We are Yankee Doodle Pilots,*
> *Yankee Doodle, do or die.*
> *Real live nieces of our Uncle Sam*
> *Born with a yearning to fly!*

At the time when Janet joined the WASP trainees acquired the equivalent of a college degree in aeronautics during their 200 hours of flight training and almost 400 hours of ground school.

Instructors were civilians, male pilots who some described as "severe and profane." Some of the men were resentful at having to teach females, grumbling that when someone asked them years later what they did in the war they would have to say "I taught women to fly." Others treated the women as they would men, cursing and yelling. Some made passes, and there were several clandestine romances. Fraternization was not officially permitted, but discreet liaisons flourished. Personality clashes were inevitable under such high stress situations, but the "Army way" of handling them was "That's tough, work it out."

Military check pilots were brought in for the official army flight tests which would determine whether a trainee went on to get her wings or, ignominiously, washed out. Nerves frayed before these grueling tests, and trainees would toss a coin into the Wishing Well and whisper a fervent little prayer to Fifinella before takeoff.

Anxiety about flight checks preoccupied the baymates. If they flunked the first test, each woman had one more chance to make good. When a trainee passed the check, she would triumphantly ring the big fire bell by the administration building so all would know the good news. Girls who washed out were crushed, knowing they had lost out on a chance to be part of something very special. They went home to families, jobs, husbands. Home to kitchens where open windows taunted them with the lure of those endless blue skies. Those who remained, rejoiced. They just wanted to fly, and now they were in the pilot seat for real.

* * *

In the mornings, after vigorous calesthenics, the women studied aeronautics, navigation, math, physics, Morse Code, and more. Physics was the only academic subject that gave Janet trouble. Interestingly, although some had worried that women would not be able to "handle the math," not one trainee washed out because of ground school.

After lunch, they took to the air. Primary training for Janet took place in the open-cockpit PT-17, a bi-wing 250 horsepower Stearman. Terrain around Sweetwater offered not only lots of open airspace for hundreds of planes but ideal ground "maps" in the layout of straight roads running for miles north and south, east and west. Navigation was by sight. At the primary level flying was taught without instruments or radios. The women practiced takeoffs and landings (always the hardest part) and acrobatic maneuvers including loops, spins, and chandelles.

Each trainee was assigned one instructor through Primary Training and different ones through the next two phases. Janet had a crush on one from Honolulu.

After two weeks of intensive instruction, the girls reached the first milestone of PT or Primary Training. It was time to solo.

Janet climbed down from the Stearman on shaky legs, grinning through the grit in her teeth, to be met by a group of baymates and friends. She knew what was coming. Laughing, the girls hustled her off to the Wishing Well and unceremoniously tossed her into the fountain, clothes and all.

"That feels great!" Janet stoutly boasted as she emerged, dripping and shivering a bit in the January air. She had been initiated.

Janet wrote home on February 3, 1944 *Pop, you can be real proud of me now. I soloed today again.* But words of pride and praise from her Father were few and far between.

Snafus were par for the course during PT. Seatbelts on the open-cockpit planes did not always function optimally. One instructor looped his plane, glanced back and found that his student was missing, seat belt left dangling. It had malfunctioned and the girl had simply fallen out. Far below, a white parachute drifted toward the earth.

During one aerial acrobatics session Janet's plane turned over and all her seat cushions fell out. Because she was short, she had secured herself in the seat with three cushions underneath her and two at her back. She found herself dangling upside down by her seatbelt, so far out that the air from the propeller was buffeting her around like a rag doll. Thankfully the plane was righted before she fell out. Janet never once had to use her parachute.

* * *

Most of their time was spent in a grueling training schedule, but the girls managed to eke out some time for fun nevertheless. Curfew was 10 PM, but the girls managed to sneak into Sweetwater on a regular basis. At the Avengerette Club at the Blue Bonnet Hotel they could drink Coca-Cola, light up a Lucky and listen to that

Big Band sound or the crooning of a young Frank Sinatra on the jukebox. Sometimes the Cokes were spiked with a little smuggled-in liquor. Back on base it was a fad to smoke sweet maple tobacco in corncob pipes after dinner.

But probably the best times of all were during those chats among baymates after lights out, when the young women could whisper into the warm darkness all that was on their minds and hearts with those who shared the same journey and truly understood. Long years later, Janet had not stopped yearning for the days of such intimacy.

Together the girls pondered their futures and wondered whether they could meet the tests that lay ahead. The threat of washing out was very real. It had happened to one of Janet's original baymates, a policewoman who had logged lots of flight hours in her civilian life. As with many experienced recruits, she had had trouble "forgetting" what she was used to and learning to fly the Army way.

More seriously, they talked about their fear of collisions, crashes, of dying. It happened. Almost every week some tragic training accident was reported.

Janet lay with her hands behind her head, staring up at the dim ceiling. "I dread tomorrow. What if I mess up again on the check flight? Twice and I'm out. I'd rather not come back at all than have to go home a washout," she whispered into the dark.

A few days earlier she had groundlooped. Coming in for a landing, she faced a crosswind, which got under the wing and lifted it up so that the other wing had scraped the ground. The narrow undercarriage on the PT 17 made such mishaps common.

"You'll do fine. Don't let it shake you up. It's happened to all of us," Patri comforted her.

"It better not happen again to me. This is the only time I can remember feeling like maybe I can't do this," Janet moaned.

"Pipe down, you guys! Some of us have to get our beauty sleep," a sleepy voice complained from the corner of the bay.

Sure enough, tomorrow was another day, and this time Janet's performance on her Army Check Ride resulted in glowing marks - all 9s.

"See, Hutch, I told you so. I didn't even think 9s existed here!" Patri gloated. "Even the Captain said it was one of the best he'd ever seen. Wait'll you tell your Dad."

Whatever the shared fears of the young women, there was a love stronger than the fear that made it all possible. They loved their country. They loved the opportunity to serve. They loved the challenge of learning, of accomplishing what no one had thought them capable of. They came to love each other. And they loved, most of all, the flying itself.

* * *

Basic Training came next. Simulated instrument training in Link Trainers (small automated mock airplane modules, complete with cockpits and instrument panels) took place inside the hangar. Like their big sisters, the link planes went up, down, and simulated the motions of flight. The closed-in Link Trainer was a sweat box. Women emerged dripping after a session to find that the 100-degree breezes outside felt pretty darn good by comparison.

Flying "under the hood" in BT-13s or "Vultie Vibrators" required a whole new orientation. The BT-13s had a canopy. Students sat in the rear hooded cockpit of a real plane, with nothing but the instrument panel visible. An instructor (or, later, a partner) sat in the front cockpit. The women were given navigation problems to solve, the goal being to end up in the right place relying on instruments alone.

Vertigo was the most difficult thing to deal with in the beginning. Repeatedly the instructors explained. "You've got to stop believing what your body's telling you and start believing your instruments instead. Your head and your stomach will tell you one thing about which way is up, and your instruments will tell you something else. Listen to your instruments." This mantra was drilled into the recruits daily.

Advanced Training took place in the AT-6 or "Sweet Six", a plane with greater power and faster landing potential. The AT-6 had a canopy overhead. Flying under instruments on night and cross-country flights were added to the repertoire. Students learned to plot their own flight plans for cross country. At night, they had to rely on their skill with the instruments, with only a small headlight lighting their way in the darkness.

Radio-beam navigation, or flying "on the beam," was another challenge. An actual beam was broadcast from station towers, and the pilot, in direct radio contact, would align herself with it for direction. As long as the student kept the plane centered in the middle of the beam, she got a steady hum. If she veered to left or right, Morse code beeping let her know she needed to readjust her course, signaling an "A" for left and a "N" for right.

* * *

Janet went on a 500-mile night solo that she always remembered vividly: the black-velvet sky, the vastness, the great silence. The feeling of flying was the closest she'd come to a spiritual experience. She reveled in the pure freedom of it all, the lift and glide into skies of infinite blue. Janet swore that the sound of the engine on a long solo flight eventually turned into music, a private symphony. *Wonder if other people ever hear that?* she wondered. Some 60 years later, she would hear almost the same words echoed by a fellow WASP alumnae at a reunion. No, she had not been alone. The celestial music had been heard by others of her sisters in flight.

Alone at night, with the stars and the moon for company on a cross-country flight, Janet felt that she had, in the words of the poet John Gillespie Magee, "...Put out my hand, and touched the face of God." Magee was an American pilot killed in action over English skies in 1941 whose sonnet glorifying flight had been discovered scribbled on the back of a letter among his personal effects.

Janet loved knowing she was *good* at flying, and other WASP veterans echoed the same feelings. The exhilaration came not only in doing what they loved to do but knowing they were gaining skills and proficiency day by day. The trainees were discovering that they could not only perform but excel at tasks men had always assumed to be beyond them. They were changing the way the world looked at feminine capabilities and limits.

And so they learned to suck it up and take it, accepting what the Army dished out in the way of grueling training and harsh working conditions in the service of a greater goal.

While the men on the front lines of battle knew they were heroes back home, the WASP put their lives on the line daily largely without acknowledgment.

Janet Lee Hutchinson, America's Youngest Active WASP
"Hutch" at age 19   1944

Janet & Dottiebell in leather flying suits
Avenger Field   1944

WASP Trainees. Janet's friend Elizabeth Erickson, right,
was killed in training mission   1944

20

THE WASP YEARS: DO OR DIE

Formation flying, by sight, was the most difficult and dangerous challenge of all. Planes flew in tight proximity and it was easy to make a mistake and clip a nearby wing.

At least one accident a day due to carelessness was par for the course. In heavy air traffic the planes had to land almost atop one another. However, the WASP had a lower average accident rate than the Air Force cadets.

When they did happen, serious crashes were devastating, both in training and later as WASP graduates were assigned to military bases around the country. The graduate pilots faced tasks as daunting as flying a war plane across the country, or providing "target practice" for inexperienced gunner trainees. Here, the WASP pilots were assigned older planes, frequently badly maintained and inadequately inspected. In simulated combat the women trailed sleeve targets for soldiers to fire at. The men used live ammo. When WASP complained about being assigned questionable aircraft, they were told outright that they were "expendable," unlike the soldiers and pilots fighting on the front lines. Male pilots themselves often refused the flying jobs assigned to the WASP as too dangerous.

The women accepted this as part of the reality of war and the price of a life in the skies. On some bases, Commanders and staff were actually defiant and undermining. Reports of sabotage, of sugar in a gas tank resulting in the death of a WASP, reached Jackie Cochran's ears. Shocked as she was, she nevertheless declined to investigate, convinced that the WASP program could afford no stigma or controversy.

Trainee and pilot deaths - 38 in all, 8 in the time Janet was at Avenger in training - were to be expected under such conditions, but the girls could not avert their deep shock when news of a fatality was announced, usually at dinnertime. To have one of their own, a young woman full of life and promise, suddenly fallen out of the skies and out of their lives, was devastating and unreal. And yet, it *was* real. Each trainee knew that it could have as easily been herself. Some accidents were particularly horrendous. One woman was killed when the hydraulic fluid tank burst inside the cockpit and flew in her face, blinding her.

* * *

It was a typical day of flight training for Janet's squad. Tired from an afternoon's maneuvers, they were coming in for landing on a crowded field. Janet's close friend Elizabeth Erickson was right in front of her in the traffic pattern. Elizabeth was a beautiful young woman, with dark hair and flashing eyes, whose classic beauty always reminded Janet of her own mother. The basic traffic pattern formed a rectangle. Elizabeth was heading downwind on the last leg of her landing sequence. Squads of upperclassmen were returning from their cross-country flights as well. Elizabeth was intently focused on her own landing procedures in a tight field, and did not heed the injunction to constantly swivel her head and scan in all directions for approaching planes.

Suddenly an upperclassman's plane materialized, heading into the pattern at a 45% angle. Neither pilot saw the other's approach as the two planes headed straight at each other.

At the last possible moment, only 800 feet above ground zero, Elizabeth managed to jump. At that height, her parachute was useless. She was dead the instant she hit the ground, her unmanned plane diving into the runway and exploding into a ball of flame. The upperclassman, too, was killed instantly.

The shocking violence of the impact, the flash of flame and the grinding disintegration of metal, the debris floating through the air, passed before Janet's disbelieving eyes like a nightmare. Somehow, functioning on automatic, she managed to bring her own plane down for a safe landing.

She couldn't look toward her friends' crumpled body. Janet walked stiffly away from the landing strip that had become, in a matter of moments, a field of death. Her numbed mind struggled to assimilate what she had seen. She moved through the rest of the day in a daze, unable to speak.

That night after lights out a whisper came from Dottiebell's bunk.

"Hutch? You haven't said a word. Are you okay?"

Janet still didn't have words for what it had done to her to see her friend go down in flames in front of her. But her bunkmate's words released, at last, the hot wash of tears that were choking her.

When a WASP was killed, she was granted no death benefits. Transportation of the body back home was not provided. The trainees themselves contributed to a fund to provide an escort home for their fallen comrade. When she was buried, no American flag was presented. Her family could not display the gold star signifying they had sacrificed a son or a daughter for their country.

Somehow the women found their own way to honor their dead, and go on.

* * *

Janet wrote her parents regularly, both separately and together. "Dear Pop" letters tended to be full of details and facts about the flying itself. "Dear Mom" letters addressed her feelings a bit more, asked her mother's advice on things and requested clothes to be sent from home. The tone of Janet's letters was generally upbeat and unfailingly loving and solicitous. It was obvious Janet still felt a great need for her parents' support and approval - especially her father's, as she strove to follow in his footsteps.

Dearest Daddy,

I got your letter the other day and honestly - Daddy - it made me the happiest girl in the world. That's the first time you've ever told me that you could trust me and I swear - pop - it really touched the spot. I've always tried so hard to make you trust me and now I've finally been rewarded...I'm trying very hard to make good - for you.

The complete trust and reliance on the infallibility of her parents had been shaped in their years together as America's Flying Family.

Although Janet was experiencing the exhilarating independence of life on her own, she still turned to her beloved mentors for guidance. Their opinion of her mattered more than anything.

Mom, I hope you don't think I've changed any. I always want to be your little girl and I'll be satisfied if I'm just half as sweet as you are.

Sometimes, frustration or exhaustion got the better of her.

I'd give anything to die right now just so I could be in a horizontal position...Things are really getting pretty rough around here. We don't kick about it much but sometimes they go just a little bit too far. I swear - a girl can only endure so much and I'm afraid some of us are going to break under the strain soon...It seems all we do all day is fly, study, and change our clothes for everything we do.

Despite her grueling schedule, as the weeks flew by Janet's confidence in herself as a pilot grew by leaps and bounds. When her mother asked her if she still got nervous about her flying performance she replied:

No. Now when I go up I just get overwhelmingly happy and know that I can do it. It's not work anymore.

Letters kept her parents close and integral to her life. Janet was used to being part of a "we" with the Flying Family, yet now she was forming a new sense of "we" with her comrades at Avenger. She sang the praises of her baymates and friends, and plotted how they could stay together after their stint ended.

* * *

When Avenger Field became the first all-female flight school, AAF cadets finishing up their training were still on base. The girls could hear the men marching, singing and laughing from across the patch of ground separating the barracks. (That tantalizing distance came to be known as "No Man's Land"). When the men finally departed and word got out about the women on base, interest ran high. Cadets from neighboring AAF training schools landed at Avenger in a continuous stream of "forced landings" for a number of days, checking things out. Soon the CO was forced to close the field to all outside air traffic.

The women were immersed in a world of double standards and conflicting priorities. Sure, they had to learn to be as tough as the guys, and then some. From somewhere deep inside, they had to muster up the downright courage and guts to fly planes that were questionable and in situations that had been rejected by male pilots as too dangerous. They had to put up with bad language, overtly condescending attitudes, and sometimes verbal abuse from their male instructors. While official heads were turned when the fellows ignored the rules, girls faced expulsion from the program for even the breath of a scandal.

Cochran's Convent, the WASP trainees dubbed their new home.

Fiercely protective of her girls, Cochran saw to it that trainees were strictly forbidden to have "social contact" with their instructors, whether civilians or AAF staff. These rules seemed destined to be stretched, bent, and broken, and they were, on a regular basis. The cooks added saltpeter to the girls' food, with no noticeable effect. Only half in jest, the girls sang:

> *Girls, girls is our middle name*
> *We are the girls of Sweetwater fame.*
> *We never neck and we never pet*
> *Give us a chance and we'll do it yet -*
> *Our instructors stay out late but we never get a date!*

The raucous marching songs the girls sang helped them blow off steam and express themselves.

> *I just called up to tell you that I'm rugged but right!*
> *A rambling woman, a gambling woman, drunk every night.*
> *A porterhouse steak three times a day for my board*
> *That's more than any decent gal in town can afford!*
> *I've got a big electric fan to keep me cool while I eat,*
> *A tall and handsome man to keep me warm while I sleep!*

I'm a rambling woman, a gambling woman and BOY am I tight!
I just called up to tell you that I'm rugged but right!
Ho-ho-ho- Rugged but right!

Disgusted, Jackie Cochran banned this one from the post. For good reasons, she was concerned about control and public image. The young women at Avenger were in the prime of their lives and at the top of their form. Their bodies were toned by constant exercise and their minds were stimulated by all they were learning. Emotions spiraled from exhilaration one moment at the thrill of living their dreams to anxiety and homesickness the next. Many were living away from home for the first time. All were caught up in the excitement and drama of a world at war. Daily, they were expected to put their lives on the line.

The atmosphere of "Eat, drink and be merry, for tomorrow you die" seemed to directly contradict the rigid disciplines of base life. Inevitably, when young hormones were raging and the heart and libido got involved, rules went by the wayside. Almost every girl dated an instructor at one time or another, meeting secretly in town or finding a secluded corner on base. Several marriages resulted from these liaisons.

* * *

An attractive young WASP was soloing in an open cockpit plane on a blistering-hot day, enjoying the combination of warm sunshine and cool breezes high above the Texas plains, when she was inspired to take off her shirt and sunbathe in privacy. Closing her eyes in blissful privacy, she was rudely jerked back to reality by the roar of another engine nearby. She found herself gazing into the laughing eyes of an AAF cadet flying just off her wingtip. His buddies soon surrounded her, waving and grinning. This so rattled her as she hightailed it back to the landing strip, fumbling to put her blouse back on, that the shirt flew out of her grasp and into the blue. Still, she managed to land successfully.

"Somebody for God's sake bring me a blanket!" she yelled, jumping out of the cockpit.

* * *

For Janet, dating had never been that much of an issue. Although she had gone out with boys in high school, she considered herself a tomboy, unlike Kathryn, "the beautiful one" who had them lining up at the door. Yet the vivacious and petite young trainee was not immune to the charms of "Steve," a handsome young civilian instructor who was wildly popular with the girls.

In the late spring of '44, Janet and Steve teamed up with two other student-instructor couples for an evening in town. So far, Janet had loved everything about her WASP adventure. Her grades were good, her test flight performances exemplary (with one or two hair-raising exceptions), and she was thrilled to be doing a good enough job to earn rare words of praise from her father. She literally couldn't wait to wake up each morning. But as the laughing couples headed toward town on that warm spring night, all that was about to come crashing down around her ears.

Sweetwater was built around an old-fashioned town square and boasted a single movie theater (or picture show, as it was then called), several churches, and one good hotel, the Blue Bonnet. While the locals had welcomed the male AAF cadets as possible marriage prospects for the girls of the town, they weren't quite sure what to make of this influx of young women. In this part of the country women in pants, to say nothing of women in cockpits, were a distinct novelty.

The park in the middle of town was a popular trysting spot. About 11 PM the young folks headed for it, bearing a bottle of brandy and a carton of Coca-Colas for setups.

"This looks like a good spot," Steve said, spreading out a blanket on the stubby grass.

Leaning back and relaxing under the stars, they had just opened the bottle, passed around the set-ups and had a first sip of their drinks when the situation began to go seriously south.

The bite of whiskey and Coke was fresh on Janet's tongue when, a drawling West Texas voice spoke from the darkness.

"Now just what the Sam Hill do you kids think you're doing out here drinkin'? This here's a public park. You're breaking town law. Or maybe military types think the law in this town don't apply to you."

A long, lanky figure in the dun-colored uniform of the local police force stepped out of the shadows. The deputy jerked his head to one side and spat a long stream of brown tobacco juice into the dry grass.

"Well I'm here to tell you that's where you made your mistake. You hotshots ain't above the law. Not in this town you ain't."

* * *

In a letter to her parents dated May 17, 1944 Janet describes what happened next.

There was nothing wrong in drinking as we are allowed to if we don't have to fly the next day. This - Mom and Dad - was the first drink I've had since I've been here - truly it was. We had no sooner had one swallow when the cops came. First they asked us if we were instructors and students and to this we answered yes. This was their first mistake as the police are supposed to have nothing to do with whether we are instructors and students or not. They told us to come along with them and when we asked them why they said it was the law that no one should be in the park after 11 o'clock...

Helpless, the young people found themselves hustled into a waiting police car and hauled off to the local jail. It was a squalid establishment complete with bare matresses, unflushed toilets and cockroaches as big as mice.

Finding herself in an actual cell rubbing elbows with a contingent of local prostitutes, Janet was stunned at first.

I don't believe I've ever been so humiliated in my life, she continued in her letter home.

The arresting officer who had been so abusive continued to harangue his prisoners, calling the girls "whores" and worse.

The treated us terrible and said horrible things to us. We kept saying yes sir and no sir as they were trying deliberately to get us mad so they could charge us with disorderly conduct. We asked them to call the field several times to tell them where we were and they said they would but they never did.

One policeman, however, was decent to them.

As soon as we were brought in he came up and spoke to us...he said that he'd gladly testify that we weren't drunk at all.

The ordeal finally came to an end about 5:00 AM.

A man from the Field came to get some other instructors out and when he saw us there he nearly died of the shock.

Knowing the police couldn't justify holding the young people merely for being in the park after 11:00 PM., the arresting officer claimed that the group had indeed been drunk and disorderly. Bailed out and brought back to Avenger, everyone was thoroughly grilled.

Later it was revealed that the hateful deputy's wife had been fired from her job at Avenger Field for a scandal with an instructor and that her husband had threatened to "get even." When information came out about the policeman's agenda and the real facts of the case, the Field got the group's fine money back and had their charges stricken off the record. The nasty cop was fired.

* * *

This wasn't the end of the matter. Base protocol and rules had been seriously breached. Within a few days a board meeting was scheduled and the girls were summarily expelled from the WASP. The instructors received a slap on the wrist with no repercussions.

While some of their superiors sympathized with their situation, no one rose to the defense of the girls. Mitigating circumstances or not, they had committed what was to Jackie Cochran the cardinal sin. They had exposed her beloved program to possible criticism and attack. Ironically, this extremely liberated woman did nothing to countermand the double standard that unfairly penalized only the females. Even the appearance of evil must be avoided at all costs.

For Janet, the worst part by far of washing out was telling her folks.

After it happened - I felt so horrible - I just couldn't call you all and tell you cause I knew it would break your heart even though I hadn't done anything wrong except break the rules--but that was enough in itself.

This was the low point in Janet's young life. For the first time, she was unable to meet a goal she had set for herself. But far worse was the fear of breaking her father's heart and seeing, in his eyes, disappointment instead of pride. The need to win her father's approval, trust, and pride had driven Janet Lee for years. Now, it seemed, she would fail.

On the day the verdict came down the effect of the washout on the young woman was devastating and complete. In despair, unable to visualize a bearable future, she headed back to the barracks alone.

* * *

Janet's legs felt like sticks of wood as somehow they carried her back to the empty barracks. She sat down on the edge of her narrow cot. Midday silence pressed in on her from all sides.

As she absently stroked the rough wool blanket and stared at the little window opposite her bed, she could hear, as if from a great distance, the swell of engines revving on the runways. High girlish laughter floated through the air, and Janet visualized the familiar scene of pilots rushing to their cockpits, eager to take off for an afternoon's practice maneuvers. This time, she would not be joining them.

The thought twisted like a knife in her gut. How could she give this up? No silver wings, no graduation ceremony, no future.

How can I tell the folks? she thought sickly. *How can I let everybody down?* The brightness of the afternoon light, the sounds of life and purpose heard through the little window seemed cruelly mocking in contrast with the breathless dimness in the room, the darkness within her mind.

I really can't imagine tomorrow, she thought dully. *I'm not a part of it anymore.*

Janet rose from her bed and moved to the windows. One by one, she closed them and pulled down the blinds. The dimness grew profound, but there was still light enough to guide her to the gas pipe by the heat register. Janet wrenched it fully open, waiting to hear the sly hiss of escaping gas begin. Then she lay down on her bunk and closed her eyes. She prayed only for sleep to stop her tortured thoughts.

* * *

An uncertain span of time later, Janet groggily swam her way back up to consciousness. She found herself in an antiseptic-smelling hospital room, tucked into a bed with scratchy white sheets. Patri, Dottibelle, bunkmates and friends surrounded her. Their faces were tight with anxiety, but she read no condemnation there. Still, Janet was ashamed. She rolled her head to one side.

"Promise you won't tell anybody!" she pleaded.

But her dear friend and surrogate Mom, Patri, determined to take matters into her own hands. She lost no time getting to a telephone and calling the Hutchinson residence in Maryland. She explained the facts of the arrest and expulsion but did not reveal Janet's suicide attempt.

"There just has to be something you can do, Colonel Hutchinson! The whole thing was so unfair, and Janet's taking it pretty hard."

The man of action wasted no time. That very day he drove to Washington DC, and confronted his old flying cohort Jackie Cochran in her office. When George had laid it all out for her, he asked if another way couldn't be found to punish the girls but avoid expulsion. The instructors, after all, had broken the same rules yet suffered no repercussions.

Jackie listened. Another hearing was scheduled and this time Janet and the others learned they would be given forty-five demerits each and confined to base until the end of training. Tough terms, but a small price to pay to keep their WASP standing.

The girls were asked if they had anything to say for themselves. Janet stood up. Her flashing blue eyes scanned the room, taking in the instructors and staff.

"Why are *we* being punished for something *you* do every day of the week?" she asked.

The room grew still enough to hear a pin drop. But the punishment stood.

* * *

The suicide attempt itself was swept under the rug in the weeks that followed. Janet never spoke of it, certainly not to her parents or even to her sister or, later, her son. In looking back on the crisis, Janet was aware that she had been functioning in a state of disassociation, so crushed by the thought of letting her parents down that she could think of only one way out. At such times (and she was to face another many years later, brought to her knees not by disappointed hopes but by the shame of alcoholism) Janet felt she was trapped inside herself, focused only on ending her humiliation and pain. Asking for help did not occur to her.

For a person who held herself to such high standards, a suicide attempt brought incredible guilt. If the burden of carrying that shame secretly and alone wore on Janet, still it was infinitely preferable to the pain of having her cherished parents know the truth. So she sucked it up, and went back to her life in the WASP with the renewed determination that second chances can bring.

Janet's confinement chafed, but she bore up under it. It was hard to see her friends, lipsticked and laughing, take off for a night in town. It helped that they always brought her treats and the latest news and gossip.

She only cheated once. On a particularly long afternoon, antsy at the physical confinement and longing for company, Janet climbed over the fence behind the barracks. She sat on a boulder and talked to a cow for over an hour. The cow ceased her grazing and stood stock still, gazing solemnly at Janet with huge brown eyes as her human companion poured her heart out. *Animals*, thought Janet, *are sometimes better than people. They always listen, and they don't tell.*

The arrival of another animal friend brought welcome diversion. Strangely enough, the WASP trainees were permitted to have pets. Several women had brought their dogs to camp, and Janet, always happiest around animals, was given a purebred white collie pup while she was at Avenger. She named the little ball of fur Eta, for the aviation term *estimated time of arrival*. Soon Eta became the spoiled mascot of the bay as well as Janet's constant companion.

Eta was the best dog Janet ever owned, and was to be with her for many years to come.

21

THE WASP YEARS: ON FOLDED WINGS

As the days leading up to graduation sped by, Janet became more confident in her skills and in herself, yet more anxious than ever that her WASP "family" would likely be scattered to the winds. There was lots of whispering after lights out about who might end up where after training ended and the newly fledged WASP took their places in the bigger world. Hoping and praying was about all a trainee could do to influence this, as assignments were apparently made without regard to merit or personal preferences.

Janet's letters home were full of reassurances to her folks that they really didn't *have* to make the trip to Sweetwater for her graduation. Katherine's wedding loomed on the horizon, absorbing energy and funds. Yet, Janet was secretly disappointed when they didn't make it for her big day.

Due to the infamous arrest in the Sweetwater park, which garnered the participants 45 demerits apiece, Janet graduated with a total of 69 demerits. 70 would have meant automatic washout. As she stepped forward to have her shiny silver wings pinned on her uniform lapel, Janet breathed a sigh of relief. *I've done it.* But it was hard to look around the crowded hall and not see her pride reflected in the shining eyes of her parents.

The hall at Avenger Field was filled with young women's voices raised in song, with heartfelt speeches, and with more than a few tears as the Class of 44-6 became pilots. Emotion ran high between a rewarding sense of gratification at all that had been accomplished and sadness at the end of a remarkable adventure. The women knew they would never be together in quite this way again. Immersed in the absorbing work that was daily life at Avenger, the girls had little awareness of what was transpiring behind the scenes as far as the future of their beloved WASP was concerned.

The Class of 44-6 went forth from their graduation ceremony expecting to be permitted to serve their country in the skies for the duration of the war. At great personal cost and effort, they had won their wings. Now, they wanted to use them.

* * *

On August 10, 1944, just days after Janet's graduation, the idea of militarization for the WASP was soundly blasted on the floor of the House. As early as March of that year hopes had soared as a Senate bill was introduced to militarize women pilots. But by April, fast on the heels of hope, the reality of the ending pilot shortage and the beginning of an anti-WASP campaign combined to form the beginning of the end.

For Janet and her friends, however, all this was still in the future. She wrote to her parents on July 9th:

Well folks - I suppose you know we're not in the army. I don't know whether I'm glad or sorry about it. Guess I'm sort of sorry. I did want to salute and wear bars. If they saw that picture called LADIES COURAGEOUS that was about us I don't blame them for turning us down.

The melodramatic film *Ladies Courageous,* starring Loretta Young as a glamorous Jacqueline Cochran, had been shown at the Sweetwater theater and energetically booed by Janet and her friends. In the film the WASP had been portrayed as flighty, man-crazy sweet young things, powdering their noses as they ran for their cockpits. None of the gritty realities of WASP life made the cut.

Regardless of what was taking place on the world stage, after graduation and months of grueling work concern about their new assignments took top priority for the new pilots. Janet was overjoyed to find that she was assigned, along with best friends and baymates Dottibelle and Patrie, to Gunter Field outside Montgomery, Alabama. Just like old times, the buddies were roommates. Dottie and Patrie were only to be there for a short time, however. Soon both were relocated to Panama City, Florida. "It felt like losing an arm to say goodbye to them," Janet recalled.

At least, Janet still had Eta, and Tommy, her new roommate. Janet's first reactions to Gunter weren't all positive. She hadn't yet encountered the type of racial prejudice that was endemic to the Deep South in the 1940s, and was shocked when a white driver forced her to move from the "colored" section in the back of the bus to the front.

At Gunter, "Hutch" was assigned to delivering planes and testing aircraft. Flying the BT-13 and AT-6, as many as ten or twelve in one day, as well as ferrying airplanes on long crosscountry flights kept her more than busy. Back on base, she also instructed.

Eta was a well-known and loved presence at Gunter as well. Out of some 300 airplanes, she could pick out the one assigned to Janet and sit on the wing, waiting for the 2:00 PM flight time. She was allowed to go along on that particular flight, as it was straight and level. When Eta couldn't go along, the faithful dog was always waiting on the sidelines.

* * *

Janet soon got a chance to practice her assertiveness. On arrival, she discovered that there was no on-base housing assigned to the WASP. The women were expected to find and pay for their own lodgings. $200 a month base pay didn't go that far, and Janet indignantly marched into the Commandant's office to demand base quarters for all WASP stationed at Gunter.

In a letter home dated 10-25-55, she explained:

We're living on the post in the BOQ (bachelor officer quarters) which naturally makes things very convenient. Mom - I just decided I must have inherited some of Daddy's "gift of gab". There is a regulation here - the only field in the U.S. - that says WASPs cannot live on the post. Well - I went to see the Colonel himself and really poured my heart out to him for close to an hour. It worked.

Janet herself marveled over her emerging role as a leader. *It seems that any time the girls need something done around here they always come to me to do the fancy talking so therefore I'm the spokesman for all of them. Doesn't sound like me does it, but so far I haven't failed them yet.*

* * *

Gunter offered far more opportunities for dating and socializing than Avenger, mostly with the young officers from nearby Maxwell

Field. Janet wrote home about a great date day she had enjoyed with a 22-year-old Lieutentant who was athletic and "not a bit mushy" (high praise from practical-minded Janet). They had played 9 holes of golf at Maxwell field (Janet beat him), swam at the Gunter pool, ate at the Officers Club, and topped off the day at the post theater seeing *Dragon Seed.*

Off on her cross country ferrying missions, Janet dashed off postcards to her folks. Exhausting as these flights were, she enjoyed seeing new places such as New Orleans and Chicago. Testing the BT-13s and AT-6s was challenging work as well. Sometimes she flew as many as ten or twelve planes in one day. Janet's skills and confidence as a pilot were growing, and were to be rewarded.

It was an honor to be chosen to take the Advanced Instruments Training course back at Sweetwater. This meant that Janet would be certified with an Instrument Rating on her license, even after the WASP disbanded. After her special training, it was back to Gunter to continue as both test pilot and instructor.

Excerpts from an article about the WASP pilots at Gunter appearing in the local Montgomery newspaper focused on Janet and her remarkable background.

Miss Hutchinson comes from a family well known in aviation circles and has been flying ever since she was old enough to sit up in an airplane. Her father, George R. Hutchinson, Ruxton, Maryland, is a pilot, author and lecturer and was once active as a trans-Atlantic ferry pilot. "Hutch", as she is known to her many friends, has flown over 300,000 miles into 41 different countries and all of the 48 states. She can keep one amused for hours with her tales of adventure as a member of the first "Flying Family" of America. She now has a total of 450 pilot hours and 4000 flying hours...Miss Hutchinson plans to make a jump in the near future together with five other WASPS, to test a new type parachute designed by her father, but the details of that venture are not yet ready for publication.

The passion Janet developed for the proposed jump to test the innovative Parabag designed by George Hutchinson reflected what came to be a near-obsession with keeping her new "family" of women pilot buddies intact in the face of the coming disintegration of the WASP.

The huge Parabag was designed to be used in group jumps during combat. Special compartments held equipment and supplies in bulk. Thus a whole squad of paratroopers could be dropped in unison, landing in the same spot, complete with whatever they needed for their mission.

Janet had conceived the idea of a test jump involving herself, Holly Stiers, and close pilot pals which would accomplish the dual missions of focusing national attention on the inventive capabilities of her father and keeping the "gang" together, and in the skies. Unfortunately, it was never to materialize.

* * *

The WASP had been informed in March of 1944 that they would soon be commissioned into the Army Air Force, and 50 WASP, including Nancy Love and Jacqueline Cochran, entered Officers Training in Orlando in preparation. At the same time, a vicious anti-WASP campaign was being waged by Drew Pearson, the noted newspaper columnist, and others in the media. Pearson demanded the deactivation of the WASP in a series of columns that attacked the organization for usurping jobs for male pilots. The Secretary of War ordered all releases about women pilots stopped while the legislation was pending, and forbade any WASP response to the attacks in print.

On April 29 the New York Daily News charged that the WASP were "jumping the gun" on Congress. The fickle public, once so responsive to publicity about the WASP, now began to view support for the WASP militarization bill as potentially disloyal to the men returning from the fields of battle.

The early fears for pilot shortages, borne out by terrible casualty rates in the RAF through 1940-41, were relieved as American casualties proved to be less than originally projected. In early 1944 the Army Air Force began cutting back on personnel, with pilot candidates transferred into the infantry. Their instructors faced going into the draft pool and eventually the walking man's army. They claimed that they needed access to other essential flying jobs such as those held by the WASP.

On a global scale nothing was more defining for American involvement in the war effort than D-Day. Just after midnight on June 6, 1944, the Allied invasion of Europe began with the landing of some 175 thousand troops at Normandy - the largest invasion force in history. General Arnold, coordinating with General Eisenhower, was completely absorbed in what would prove to be the beginning of the end of the war in Europe

As the pilot shortage was ending in the final months of the war, civilian male pilots formed a lobby to attack the WASP militarization bill.

On June 21, the hearing on HR 4219, the bill calling for militarization of the WASP, lasted all of one hour. Supporters faced a bitter, vocal lobby of thousands of out-or-work male pilots and veteran's groups. The bill was defeated despite the support of the President, the Secretary of War, and the Commanding General of the Army Air Force. On June 26 the House recommended discontinuing the WASP training program and Arnold ordered the WASP officially discontinued as of December, 1944.

* * *

In early June, the Washington *Post* proclaimed: "The House will face a battle of the sexes when it considers the bill to militarize the WASP." Truer words were never spoken.

On August 10, 1944, just days after Janet's graduation, the idea of militarization for the WASP was soundly blasted on the floor of the House. Snippets of the dialogue and diatribe vocalized during the hearings exemplify what the WASP faced in Congress and on a national front:

Congressman James Morrison of Louisiana, one of the fiercest opponents of the WASP bill, drawled that the WASP was "the most super-duper and glamorous of all programs", and described the WASP uniforms as "tailored on Fifth Avenue in New York and cost over $500."

Charles Elston of Ohio: "The women who are today piloting all kinds of planes across the country are rendering a magnificent service to the country. They should be accorded all of the benefits of a soldier and are entitled to be commissioned as provided for in this bill."

Joseph P. O'Hara of Minnesota worried that for the military to "reject" less-qualified male pilots in favor of more-qualified WASP would mean "...that boy must go back and become a grease monkey or tail gunner...because his natural flying ability does not fit into this program...because they say, 'We have to have somebody in there who is a very attractive lady pilot.'...Why this is a piece of social legislation in my opinion and that is all it is."

A colleague: "I think it is time to forget the glamour of this war and think more of the gore of war."

Many familiar names are to be found in the minority who voted against killing the WASP bill: Lyndon Joyhnson, Clare Booth Luce, Estes Kefauver. The bill was lost by only 19 votes.

These legislators, whether bitter and bombastic or supportive, still were operating within a worldview which saw women and their place in the world as fundamentally and forever different from that of men. Through hard, gritty work and self sacrifice the WASP had won for themselves, for a time, independence and respect. They would not be permitted to keep it, for society and the powers that be did not regard it as their birthright but as a loan to be called in, with interest, on demand.

* * *

More than a little late, Jacqueline Cochran went on the defensive with an official report on the complete history of the WASP program and her recommendations for its future. The ban on WASP publicity was lifted by the War Department, and on August 7 Cochran's report was released to the media in its entirety.

Here it must be said that Cochran's agenda and rigidity in maintaining her singular vision for the WASP can only be seen as factors hastening the program's demise. She summarily dismissed the possibility of incorporating the WASP into the WAC. Unlike Nancy Love, who was willing to accept a more limited role for her ferrying squadron, Cochran insisted that the WASP *not* be assimilated into the WAC, and that if they were not to be militarized under her own command, that they be disbanded entirely. The Air Transport Command, under Love's supervision, was not consulted before the Costello bill went to Congress, and the Ferry Division, which utilized WASP personnel most intensely, was never called to testify.

In her final report to Congress on the WASP, Cochran concurred in the call for deactivation:

An inspection trip which the Director of Women Pilots took in the late summer of 1944 to more than 50 bases where WASP were employed, and discussions with commanding officers, convinced her of the correctness of the view of the Assistant chief of Air Staff...who had recommended that the WASP be inactivated. She therefore concurred in the recommendation.

Still, the report's conclusions made some strong fact-based claims as to the validity of the WASP program. Statistics didn't lie. Testing and extensive medical research had also helped clarify women's ability to do the job. Among the findings:

Women...can be trained approximately as quickly and as economically as men in the same age group, to fly all types of planes safely, efficiently, and regularly...Physiology peculiar to women is not a handicap to flying or dependable performance of duty...The flying safety record of women pilots approximates that of male pilots in the same type or work...Women pilots have as much stamina and endurance and are no more subject to operational or flying fatigue than male pilots doing similar work...

Cochran's report concluded with a strong statement about the impact of the failure of militarization itself on the individual WASP:

The failure of militarization may not have shortened the life of the women pilot program, for the situation with respect to available pilots rapidly changed soon thereafter, but it left the WASP on deactivation without any rights or veterans benefits; it left them without reserve status that might otherwise have been possible, and even desirable from some angles; it left the next of kin of those who died in the service without any insurance, and even without the right to display the gold star.

The ever-fickle American public had greeted the upbeat stories of cute WASP pilots in pigtails working on a fuselage, attractive and vibrant young women marching smartly in their sharp new uniforms or lounging on Avenger Beach, with the kind of indulgent interest they afforded to Hollywood glamour gals. And, in all fairness, for a time there were stories that recognized the WASP for their true accomplishments and patriotic zeal. But when the tide turned against the program, this type of superficial support was not enough to keep the WASP program aloft. Real patriotism meant supporting our boys on the front lines and elsewhere. The "girls," whatever their contributions and sacrifice, became a footnote in the

American wartime consciousness. After the war-torn Forties receded into the bland Fifties, even that dim memory would fade.

* * *

The graduation of the last class at Avenger Field, 44-W-10, was held on December 7, 1944. It had been three years to the day since Pearl Harbor changed the world forever. Sweetwater alumnae returned "home" for the last, bittersweet ceremony. It was attended by four generals, including Hap Arnold, and of course Jackie Cochran. Nancy Harkness Love had answered her written invitation succinctly: *No answer as far as I'm concerned, NHL.*

Letters Janet wrote home during this period reflect her growing preoccupation with the future, not only her own but that of all her sisters who had been awarded wings only to face having them ripped away. Although the WASP themselves were forbidden to contact their representatives directly to solicit support for the militarization bill, many parents did so in their stead. Janet constantly importuned her father to intervene, to use his influence in behalf of his daughter and her friends, if not for the WASP as a whole. She poured her heart out on the subject in a letter to Blanche:

You see - I'm just like Daddy - now that I've had this wonderful training I know I'll get restless if I stay at home and be just like every other girl. I've simply got to do something different. Not just instruct at some little airport, but something bigger. Surely Daddy can do something for me within three months - and Mom not only for me, but for my friends, at least one of them anyway if worst comes to worst. Just to have a companion that can speak my kind of language would be all I'd want. I know that some things are out of the question until after the war, but still - there must be something somewhere I could do. If we and the WASP could only organize again with some big company...

Janet's fertile imagination ranged far and wide in conceiving of possible futures. She, Tommy and a few others actually volunteered their services to China, the country being in dire need of pilots in their fight against Japan. Janet's parents recoiled at the idea of their daughter in hostile skies half a world away. In letters, she debated with them, trying to get them to see it her way:

China has taken every pilot she's been able to get her hands on - she's in desperate need of them. If we were accepted I know it would be the worst place in the world to go, but then we can help a little and that's all that really counts...China is so bad off. I for one - and I'm sure the majority of the other WASPs feel the same - would be glad to take that chance in order to help them. I know we could do a good job and if it did so happen that we went - why - we'd be known everywhere for our courageous work and after the war was over we could really feel as though we had contributed something. Mom - you know yourself that it is only right to put yourself out and risk your life maybe in order to help the underdog...try to put aside the fact that because I am your daughter you naturally want to protect me and keep me out of danger. I'm not afraid and I never will be.

Such idealism and courage would go begging due to the rejection, in the months ahead, of any and all WASP offers to serve their country in the skies. Even an offer to continue flying WASP missions free, entirely on a volunteer basis, was summarily dismissed.

Janet's creative mind conceived of an airline run entirely by women. To her father she wrote *I don't see how it could possibly go wrong as transportation is so in demand nowdays.* She wanted him to search out a wealthy backer as he had done so persistently and effectively in the past, this time promoting not just his own endeavors but hers. *Surely you must know somebody,* she implored.

Half in earnest, Janet wrote:

I thought of the funniest publicity stunt last night. I was thinking if we ever started a women's airline we would probably have a great deal of difficulty getting people to be our passengers at first being that we were women...wouldn't it be wonderful if for one year we could keep it a secret and not let anyone know that women were doing all the piloting. It could very easily be done you know as the passengers would only see the back of their neck - and then after we've established a good record we could publish the fact that women were flying them all the time, I bet you anything that that would get rid of all their feeling toward us. It probably sounds awful silly Pop, but I bet that if it could be done it would really go over great!

* * *

There were dark times, too, still to be endured.

I just got word from Tommy that Kay Dusack was killed yesterday. It happened when she was flying on instruments from Washington to Cincinnati. She was the head of all three training commands and had done her job well. Oh gosh - Pop - I get so sick of having people go like that. Only last week she was here and we had a nice little chat. Tommy is taking it rather hard as she knew her very well. For the last hour or so we've been all sitting here in the living room, talking about the deep subject of death. Consequently at the moment I feel very depressed but it will pass I suppose...

At these times Janet's longing for her parents became intense.

Right now if one wish were to be granted to me I'd wish that you and Mom could be here with me now...no matter how many wonderful people you meet of how many close friends you have they just don't ever take the place of your parents. No matter what I'm doing or where I am I always wish you were there with me.

* * *

On October 2, 1944, letters were sent to all active WASP announcing Deactivation as of December 20. Finally, the other shoe had dropped. The cut-and-dried directive by Cochran outlined details of the discharge process. General Arnold added words of appreciation and farewell.

To the women whose lives had been, for two years, utterly dedicated to the WASP enterprise, deactivation came as a shattering blow. They were being taken out of the game while it was still in progress, and would not be on the playing field when victory came.

Their feelings are perhaps best summed up in a final song:
We wanted wings, then we got those goldarned things
They just darned near killed us
That's for sure.
They taught us how to fly
Now they send us home to cry
'Cause they don't want us anymore.
We earned our wings, now they'll clip the goldarned things.
How will they ever win this war?

Nancy Harkness Love summed up the unique situation of the WASP with these words: *Ours was the only women's service which was subjected, in the line of duty, to the same hazards as men.* As a group the WASP had flown over sixty million miles. The thirty-eight women who had given their lives for their country, flag or no flag, gold star or no gold star, would remain unsung heroines. Thirty-four years would pass before women would pilot military planes once more.

Submitting reports and documentation of the WASP program to the United States government following deactivation was akin to tossing information down a black hole. Sealed in government archives as classified information, the WASP history was forgotten for many decades. It was as though the Air Force and America itself completely forgot that women had ever flown heavy bombers, exposed themselves to live amunition fire or died in the service of their country.

* * *

World War II would drag on for nine more months. For the women who returned sadly home for Christmas of 1944, another kind of war had ended - not with a bang, but a whimper.

Janet's letters home during this time reflect the frustration that ran deep as all WASP faced their uncertain futures.

I hate to say this, Daddy, but several times during the past few weeks I've thought of going straight home and not waiting for December 20th. In fact I even went so far as to start packing one night. I know it's my duty to stay until then and I will - but Pop - it's hard cause I get so very restless thinking of all the things I want to do when I get home. My hands feel as though they're tied behind my back here...All I do is fly all day and then go home and write letters, wash Eta, study, or anything to keep my mind off other things. Once in a while I take in the show on the post. Usually only make a date about twice a week."

For the usually exuberant Janet Lee, the dispirited tone was obvious. Still, she laid her plans and when December 20 finally rolled around, she wrote to her mother.

Yes Mom, today's the big day. The day that I hoped and prayed would never come but now that it's here I suppose I'd better make the best of it. Seeing you will make me feel a lot better.

To postpone the wrenching sense of separation Janet invited her five baymates to the Hutchinson's Ruxton home for a few days' R & R. The girls bunked down in the dining room, feeling like roomies once again. The group caught each other up on all their doings since graduation, retold old stories, shared their hopes and fears for their futures. Like schoolgirls they laughed and whispered into the wee hours. When would they ever again be with comrades who understood so well and without words the tremendous experience they had shared?

A road trip seemed like the perfect activity to commemorate the end of their adventure. The six friends piled into a coupe, packed snug as sardines, and took off for California. The girls took two-hour stints at the wheel as they traversed by car the vast country they had flown over in war planes just weeks ago. They stopped off to gape at the Grand Canyon, then found their way to a log cabin in the mountains around Lake Arrowhead. For six blissful weeks they enjoyed days in the sunshine and fresh air, nights by a fire underneath the incredibly bright stars. They played cards, drank, smoked, giggled, and cherished friendships still held close. Truly these were the best of times.

The road finally ended in Hollywood. Janet was back in the exciting world she had left behind years ago. Playing tourist, she took in the sights with her girlfriends. As time for parting approached things got a bit frenetic. The friends held a contest to see who could down the most shots of beer in a minute flat. Janet came in third. At this point, drinking for her was still social and fun.

But all good things come to an end. The jaunt finally over, the six friends dispersed to their old homes and new lives.

* * *

For the returning WASP, as for the soldiers returning from the theater of war, it was a challenge to adjust to life on a smaller stage. But re-entry for the women pilots brought additional difficulties.

The GI Bill allowed thousands of returning veterans to attend college and thus begin to build lives full of new potential. This opportunity was denied to the WASP. Looking for jobs in the private sector as engineering test pilots, interviewing with national aircraft companies or airlines, produced little more than offers to become stewardesses. Girlfriends who had stayed at home often were better trained, due to their wartime office or factory work, to to assume what few positions were available to women in a post-war marketplace.

Because the WASP vanished from the public eye so completely and so long, a special kind of isolation descended on the women involved. The WASP saga did not appear in history books, was not taught in schools. Their contribution was not recognized or validated. Often even their own families and children knew little or nothing of what Mom had done in the war.

An attempt was made to keep the WASP connected, if not to the national consciousness, then at least to each other. The Order of Fifinella was formed at Janet's base, Maxwell Field, just prior to deactivation. In 1947, a gathering of former WASP in Washington, DC was attended by only one participant - the organizer. The women had become immersed in marriage, motherhood, jobs. And for many the loss was still raw.

It would be 34 years before military recognition of the WASP was pushed through Congress by a bill sponsored by Senator Barry Goldwater and supported by Janet Reno, Jacqueline Cochran, and Bruce Arnold, son of General Hap. Under it, the WASP were at last granted full military status and were eligible for veterans' benefits. While opposed by the Veterans' Administration and the American Legion, and after a terrific behind-the-scenes fight, the bill was finally passed on the floor of the House on November 3, 1977 and signed into law by President Jimmy Carter. It was a new world order for women. At last the contributions of the WASP would be honored and celebrated.

Looking back on the WASP saga almost 60 years later, an older Janet sighed.

"What people don't think about is that when the last WASP goes, that's it. There'll be more enrollees coming along in every other branch of the Armed Forces. But for the WASP, when the last of us dies, that's the end of the story."

Her clear blue eyes twinkled. "You know, I was the youngest coming in. That should mean I get to be the last to go!"

22

HOME TURF

For Janet, the return to Maryland and "normal" life was a mixed bag. She was back on home turf, back with beloved family and friends, and, of course, her horse Laudy. But what now? What was a WASP with clipped wings to do with the rest of her life?

Having grown used to such intimacy with her baymates and friends in the WASP, it was difficult to reconnect in everyday life. Relationships which had been close in the past became strained and distant. What did she have in common with girlfriends who had stayed home? Like returning veterans everywhere, ex-WASP found that the enormity of the experiences they had shared created a gap with "civilians." Assimilation would take a while.

In 1945 it was very much still a man's world and Janet was only 19 years old. In her attempt to emulate her father, to grasp her destiny and make it her own, the young woman had, for a golden time, soared. Now she faced an uncertain future, but still looked skyward to plot her future course.

For a time, Janet relocated to McLean, Virginia and flew for the newly formed Reconstruction Finance Corporation, a private enterprise that sold and delivered no longer needed military aircraft to private industries and individuals. Planes were a surplus commodity in those postwar months and were offered "as is", without repairs and frequently missing important equipment such as an altimeter or air speed indicator. Highest bidders got bargain-basement prices.

Janet only flew single-engine planes in this job, delivering them to points across the United States. Sometimes the aircraft were in spotty condition. Moselle Simpson, a WASP upperclassman who Janet had known at Sweetwater, worked for the same company. One day Janet watched Moselle's plane, flying directly in front of her own craft, sputter, fail, and crash into the ground. Moselle was injured, but survived. Janet visited her in hospital, but didn't allow her friend's accident to shake her too badly. At least she was still flying, which was more than could be said for the vast majority of disenfranchised WASP.

The boss, a jovial married gentleman, joked with his petite blond pilot that he could easily be arrested for transporting her across state lines. Part of her job was to co-pilot a Lockheed Electra, flying company VIPs to their meetings in other states. The meetings lasted, perhaps, 45 minutes. The executives occupied themselves on board the rest of the time drinking and partying.

They met with such bad weather on one trip that Janet was ordered to hold in an upper pattern over the airport for over an hour. The men, all aged fifty to sixty and extremely well lubricated, became almost hysterical with fear. Some cried, believing their time was nigh. Janet's boss sent her back to reassure them, which she proceeded to do most effectively.

"It helped that I still really didn't relate to physical fear. I'd been in many far worse situations," Janet recollected with a smile. "Those men - such giant egos, but when they got scared, they reminded me of babies."

The job eventually came to an end, for she had only been offered the position on a temporary basis until the young man who had previously held the job returned from the war. Janet returned to home base with her parents in Maryland, and began instructing at local airports.

She started out at Glen Burney Airport, but it was such a long commute she soon switched to Rutherford, a short distance from Woodlawn and her home. Her students were 100 % male. The steely nerve and easy confidence of their young instructor usually inspired confidence, but occasionally a novice student got shaken up.

On one training flight the Cessna's propeller abruptly stalled. Keeping her cool, Janet brought her down for a forced landing, just barely making the runway. Her big, burly male student jumped out of the plane as it touched down and promptly fainted dead away on the tarmac.

Janet did some modeling and enjoyed a full social life. Her social group was extremely close. The same five boys and five girls from high school still dated within the group, interchanging partners frequently. Eventually, for Janet, the field narrowed all the way down to Bob Simpson, a boy she had met at a McDonough Military School dance. Although it was far from love at first or even tenth sight, still all the pieces seemed to fit for the young couple. Her friends were pairing up and settling down. Janet found herself slipping into the expected role for a young woman in those postwar years. Marriage and family loomed on the horizon.

* * *

Janet had always seen herself as the tomboy, while Kathryn was the boy magnet. But she was lovely and vivacious, with curly honey-blond hair and a million-dollar smile. Her heart had never really been stirred, and she remained deeply naive about sex and romance. Was this really love? Was this right? Everyone around her, and especially both families, seemed to think so. Uncertainty was brushed aside in a forward momentum of courtship, engagement, and wedding planning as events seemed to take on a momentum of their own.

The Simpsons and the Hutchinsons were delighted with their offspring's plans and heartily approved the match. Bob's family was moneyed and socially prominent. The young man himself was handsome and charming, with good prospects. True, Janet had a problem with her prospective mother-in-law's bossiness and controlling nature, but then so did everyone else. In contrast she found her new father-in-law "a sweetheart."

At Janet's wedding shower, her giggling bridesmaids gave her a nightgown that was sewn shut at the bottom. Confused, Janet asked for scissors to cut it open, provoking even more laughter.

Blanche had been reticent and embarrassed, in the way of the times, in explaining the facts of life. Her only reference to sex consisted of presenting her daughters with a copy of *Marjorie May's Twelfth Birthday,* which was supposed to answer all their questions. It didn't.

* * *

Janet was 21 years old in 1946 when she married Bob Simpson. The wedding was traditional, with all the trimmings. Mary, Janet's WASP roommate, was her maid of honor. The "group of five" baymates were bridesmaids. A glowing young bride waltzed down the aisle on her proud father's arm. But in her heart, doubt and uncertainty reigned.

"I could see the same doubt in Bob's eyes," Janet recalled. "I don't think either one of us knew what we were in for."

Certainly, the honeymoon itself cast an early shadow on the prospects of "happily ever after." The newlyweds honeymooned at Sherwood Forrest, a resort where the Simpsons owned a log cabin right next to the golf course. On the second night, Bob's brother showed up at their door. The two men took to the fairways, leaving the new bride to fume and pout. Bob came home one night from drinking with the boys to find his golf clubs in his bed, instead of his wife.

Still, Janet was prepared to make the best of things. She was married, and what was done was done. The years that followed were to be strangely emotionless in her recollection as she went through the motions expected of a young wife. What was missing was the intensity of passionate involvement she had had for a time, in the WASP.

Janet continued instructing while putting Bob through his last year of college at Johns Hopkins. For a time, the couple lived with the Hutchinsons in their large Maryland home. Janet gave up flying and took a job on the night shift at Shepard Pratt, a renowned Baltimore psychiatric hospital. One night, she was instructed to deliver a message to the third floor, the ward where the sickest patients were housed. A female patient turned on her viciously and spat on her.

Since they had no car, Janet walked the two miles home. She had a standing arrangement with Bob that he would meet her halfway at midnight. He never kept that agreement.

For the young couple, parties and drinking were a normal part of their social world. Returning home from a party where she had drank too much, Janet collapsed on the bed, stomach heaving. Suddenly she jumped out of bed, ran to the window, and stuck her head out. Wretchedly, she threw up onto the porch roof. Anything was better than having her state exposed to her parents, who occupied the bedroom right next door. Fortunately, it rained that night.

In a few months Bob and Janet acquired their own automobile and moved into their first ground-floor apartment in a building nearby. Janet still had her collie Eta, who was supposed to faithfully guard the premises. A thief struck one evening, however, ransacking the apartment and stealing their car (which was later found abandoned).

Domesticity didn't come naturally to Janet, and cooking was never a priority for the girl who had happily shared a lion's raw-meat diet. Bob invited a friend home for dinner one night, and Janet served pork chops.

Her husband glared at her from across the table: "Pork isn't supposed to be *gray*," he hissed. Angrily, Janet scraped the remains into the garbage.

She darned his socks only once. He got a blister. Life, like the pork chops, frequently seemed gray. Compared to the intense sense of aliveness and involvement she had found in the WASP, real life sometimes seemed to be a washed-out imitation.

* * *

Janet changed jobs frequently. For a time both Simpsons were employed back at Bendix, Janet then worked as a clerk-typist in the court house in Towson. Her boss became her best friend. Doris lived in the same apartment complex, and like Janet was an excellent and avid golfer. Frequently the two would head for the Hillandale Country Club driving range, sit and drink the half-pint of booze they had brought with them, then hit balls. The two athletic friends alternated winning the club's golf championships. Both loved to sing, harmonizing around the piano at club functions. This attention brought them free drinks from impressed guests. The two remained close even after the government downsized and let a group of typists go. Janet, the newest and slowest, was let go first.

She began to long for a baby. Kathryn already had two young children, and it felt like it was time. Bob, for whatever reasons, was at first uninterested in fatherhood. When Janet complained to her doctor, he suggested she purposely start "forgetting" her diaphragm. Naively, she believed that the very first time without protection she would automatically get pregnant. Instead, it took some six months of conscientious "forgetting" for conception to occur.

Janet found that pregnancy (second only to her time in the WASP) was the happiest time of her young life. She felt fit and healthy, continuing her active lifestyle right up to the end. Janet remained so svelte that she never had to wear maternity clothes, but simply let out a few seams in her skirts. And if Bob seemed distant and uninvolved, Janet just shrugged and turned her attention to the excitement of the coming baby.

If Janet's pregnancy was full of joyful expectancy, the birth experience itself was a nightmare of the first order.

23

BEGINNINGS AND ENDINGS

Janet was swimming at the country club pool when her water broke six weeks ahead of her due date. Blanche blamed the fact that her willful daughter had been horseback riding just the day before. A girlfriend drove a dazed Janet home, and Bob took her to the hospital. The intensely painful labor went on until 10:00 PM. The baby was in "Frank Breech" position, emerging rear first. Ordinarily the doctors would have tried to turn it, but due to the prematurity things were happening too fast. Today, Janet would have been given a Caesarian. As it was, she felt she was being torn apart.

The pain became an entity, picking up her small frame and savaging it in merciless jaws. At one point she raised a sweat-soaked head from the pillow.

"What animal is that screaming?" she cried. "Help it!"

"Shhh," whispered the head nurse. "It's almost over," she said, as she reached out to press Janet's straining body back to the mattress. At that moment she realized who was screaming.

It's me. The animal is me.

The doctors refused to give Janet any medication, insisting she had to be alert to keep pushing. Desperation giving her strength, Janet fought her way out of bed and tried to run to the window and jump out. The nurses stopped her.

* * *

When the agony was finally over and the nurse put the tiny, ghost-white form into arms that were almost too weak to hold him, Janet gaped at her little son in disbelief. Her baby was shriveled and beaten up from his birth ordeal and looked more like an ancient old man than a newborn. His legs were bent back under him from his breech position. It would take them several weeks to straighten out.

Little Robert, called Robbie, was kept in the hospital in an incubator for a week after Janet was allowed to go home. His mother was in a state of high anxiety. Would her baby ever look normal? Had his brain been affected by his birth ordeal?

In the course of three months, Robbie grew into a healthy and beautiful baby. The sun came back out. Janet exulted in motherhood and the new life that revolved around her son. Still, she had made a vow on that day of pain that she would never have another child. She kept that vow.

* * *

Once his son was born, Bob made an abrupt about-face and became a devoted father. He even changed diapers. As time went on he was gone more frequently on real estate deals that took him to New York and further afield. Janet fell back on Doris for companionship, especially after the little family moved to her best friend's apartment building. They were right downstairs from Doris and her husband. Doris became a second mother to Robbie, and was she who was there there when Robbie took his first step, toddling across the kitchen floor from his mother's arms to hers.

Eta, Janet's collie, was still very much on the scene and adored Robbie. She wasn't too thrilled, however, with babysitting detail. Janet would put Robbie in his bouncing chair in the yard and instruct the dog, "Eta, *stay!*" and go shopping for as much as an hour, confident the dog would guard her son with her life. Eta hated being left behind, and her reproachful eyes would follow Janet down the street.

Sexual incompatibility was another cause of marital unrest. Bob's proclivities for having sex in unusual places - the closet of the apartment, even the golf green at the country club - seemed strange to Janet. The more she withdrew from her husband, the more he shut down to her.

When Bob was out of town, Janet and Doris got together frequently in the evening for a few drinks. Gradually, Janet and Bob grew further apart. Although they seldom fought, and even though

Bob adored his little son and was an attentive father, the bond between husband and wife simply didn't exist.

The disintegration was slow at first. If Bob told Janet he was in New York on business, she believed him, though his business trips grew lengthier. Janet kept her head down, concentrated on raising her son, and made the best of things. Inevitably, the charade of a happy marriage was revealed as a sham.

* * *

One evening Janet and Bob were in the hall of their apartment, watching their eighteen-month-old son in the bathroom, when out of nowhere Bob blurted out the words that would change their lives.

"I want a divorce."

Janet turned a stunned face to her husband. The words that followed cut even deeper.

"I just don't love you anymore. I can't go on being a hypocrite."

In the stilted conversation that followed the truth came out. Bob had fallen in love with another woman, a nurse, and wanted to marry her.

In the days that followed Janet struggled to take it all in. How had she been so blind to the infidelity? Had she really known this man at all?

Bob moved out immediately. For a time, Janet seemed to function quite well. She even got together with Bob and his new love for dinner to discuss the future.

"We all sat there chatting, just like old friends," she recalled. "It didn't seem strange at the time."

Reality finally penetrated the numb haze that surrounded Janet when Bob remarried the day after the divorce became final. Finally, the pain began to surface. Janet ran from it with the same kind of unthinking desperation she had experienced when threatened by expulsion from the WASP. Grabbing the only thing she could put her hands on that might do the job, she downed a bottle of iodine.

* * *

Her throat was on fire. It felt like strong hands were around her neck, choking off breath. All of a sudden, she remembered little Robbie asleep in the adjoining nursery.

God, what have I done? she thought, grabbing the phone with shaking hands and dialing Doris's number upstairs.

Sirens. Police. Glaring lights. The humiliation of being hauled into the emergency room, having a tube jammed down her raw throat, having her stomach pumped. Janet had been here before, but then she had been a a young girl, not a single mother totally responsible for the wellbeing of her little boy.

Hollow-eyed, she woke to face the new day. Again, she felt, she had failed miserably. Huge questions about the future loomed.

How can I raise a child all alone? And what am I supposed to do with the rest of my life?

A kind of paralysis of the mind and will left Janet feeling washed out and without energy or direction. She sent Robbie to her parents' house and spent six weeks at her uncle's, lying on the couch. She saw a psychiatrist once. It was time itself that seemed to help as she began to reemerge from her protective cocoon. Janet got off the couch and back into her life.

Doris was a mainstay through all the chaos. The girls went out together several evenings a week, and sometimes would be joined in their drinking parties by Kay, a friend who worked in a medical research lab. Soon, Janet became sickened by Kay's gory tales of what happened to the research animals. Still the ubiquitous Kay insinuated herself into their lives, flirting outrageously with any man in sight, including Janet's post-divorce dates.

Janet had had to shop for a used car after Bob left, and the salesman who helped her out became infatuated with her. "Hank" insisted on giving her the car outright. It was a bullet-bodied Hudson, complete with rumble seat. Hank hung around the apartment, mooning over Janet, but she rejected him as off-limits (he was a married Catholic). Still, the attention was flattering.

She was shocked one evening to come home and find Hank and Kay entwined in her living room in a very compromising position. Later, she spied unquenchable Kay yet again in the parking lot, making out with Doris's husband in his car. Doris was livid. That night it was impossible for Janet to tune out the shouts and thumps coming from the apartment overhead.

* * *

Kathryn's life, in constrast to Janet's, had proceeded along a fairly straightforward track. After graduating with a major in English from the Maryland College for Women, Kathryn married

Hillis Hume in the Episcopal Cathedral in Jacksonville, Florida in 1944. Hume was a graduate of the Naval Academy and had been an All-American football hero.

As a military wife Kathryn moved from post to post in Florida cities as her husband underwent flight training. Hillis went on to become a Lieutenant Commander, a weather pilot who flew into the eyes of hurricanes to assess developing conditions.

At the time of Janet's divorce Hillis was stationed in Trinidad, where he and Kathryn enjoyed the country club lifestyle. It was a perfect time for Janet to visit and truly get away from it all. At their urging she flew down for almost a month. What with golf every day, lunch at the club, drinks and laughs around the pool, new people and faces, Janet suddenly felt young and alive once more.

Then she met Norman, a handsome enlisted man who was also a golf pro at the country club. Norman was devastatingly popular with the ladies. Janet cast off her memories of a failed marriage along with her self-restraint as the couple partied, drank, danced and laughed.

Kathryn, however, was beside herself. Naval protocol forbade officers' family members from fraternizing with the enlisted men, and Kathryn tried to stop her little sister from getting involved with Norman. She had little success. Janet snuck out to an overnight party which turned into a strip poker game. Kathryn was fuming when an unrepentant Janet crept back into the house the next morning.

Janet was infatuated, but wary, and she resisted Norman's amorous overtures. Stimulated by the challenge, he soon followed after when Janet returned home to the States.

Norman stayed with Janet's neighbor Esther, Blanche's best friend. Robbie adored him, and it seemed to be mutual. It didn't take Norman long to propose. Janet accepted.

The engagement was destined to be short-lived. Returning to duty in Trinidad, Norman was caught sleeping with his C.O.'s wife and was thrown in the brig. Sheepishly, he called Janet and broke it off.

Janet realized from her lack of reaction that she hadn't really loved Norman. He'd been a rebound relationship for her, and to him she'd simply been the girl of the moment.

* * *

Back home, Janet still felt the need to reach out for some kind of anchor. She became involved in a Baptist church community headed by a dynamic, almost hypnotic young minister. He showed a real interest in the pretty young divorcee. But it was not only in saving her soul. Visiting Janet in her apartment, he suggested they get down on their knees to pray. When she dutifully closed her eyes, he grabbed her. She fought him off without much trouble.

After this fiasco, man trouble receded into the background for a long while. Facing life as a young single mother, it was time for a serious attempt at reconstructing a life.

Janet and Robert Simpson   Wedding Day August 1, 1946

Janet and 18 month old son Robbie   1952

Janet and Kiddie Kollege Graduating Class   1975

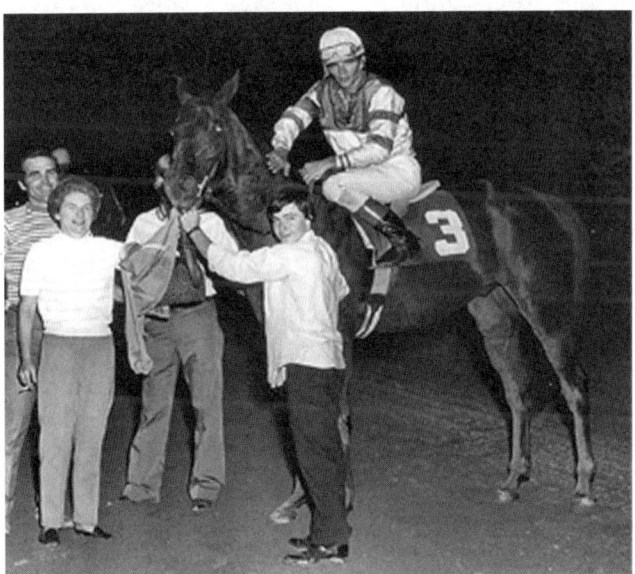
Owner-Trainer Janet with beloved racehorse Brasher Falls

24

KIDDIE KOLLEGE

A return to flight instruction was the most remunerative and fulfilling thing Janet could conceive of doing at this time. Accordingly, she resumed her instructor's role at Rutherford Airport. Blanche looked after Robbie during the day. It was good to be back in the cockpit, but Janet hated to be away from her son, returning home after a long day and commute only to find Robbie already in bed fast asleep.

It was Blanche who suggested in 1953 that Janet start a nursery school so she could be home with Robbie during the day. The Hutchinson estate provided a perfect setting. On the grounds there was a well-built structure that had, at one time, housed some 200 chickens and exotic fowl. It was still home to peacocks, guinea hens and other exotic species. The "chicken coop," remodeled, became the start of Kiddie Kollege, an enterprise that would occupy Janet for the next 39 years of her life and become immensely successful.

Beginnings were modest. Janet started with just five children ages three through kindergarten, one of them her own. School met from 9 AM to noon.

Kiddie Kollege was an idea whose time had come. In the early Fifties the whole concept of day care was brand new, even in Ruxton, one of the most prestigious and wealthy Baltimore suburbs.

"We were the only family in the neighborhood who weren't rich," claimed Janet.

Applying many of the principles she and Kathryn had become familiar with in their "Classroom in the Skies" Janet offered, in Kiddie Kollege, a combination of high standards of learning, individualized care, and just plain fun that made the institution a success from Day One. It didn't take long for word of mouth to spread. As the program expanded, Kiddie Kollege became *the* place for upwardly-mobile suburbanites to place their little ones. Janet went back to school at night to earn her teaching credentials and ran Kiddie Kollege during the day. Her friend, fellow teacher and RN Bettemae took classes alongside her. Janet's days were incredibly full, as she build success in a whole new arena. Best of all, she got to be at home with Robbie. Life settled into a comfortable routine.

* * *

For a time, George Hutchinson watched the school grow from the sidelines. Although he continued to write and to make occasional public appearances and speeches, still he was unfulfilled at his restless core. In the years since Flying Family fame had waned, George, whose mind and will were as strong as ever, had tried repeatedly and mostly in vain to regain the limelight he had once enjoyed.

In the early days of WWII, a speech he delivered on *Can the United States Be Bombed?* in New York City had drawn a standing-room-only crowd. The speech proved to be prophetic, borne out by the Pearl Harbor attack, only months away. George wanted to use this same visonary and creative capability to go on capturing world attention, but also to help change the world for the better. He was a true idealist.

Driven by a kind of patriotic altruism, George wanted, most of all, to literally change the way the government and the military looked at war, defense, and peace. Accordingly, he turned his considerable energies to inventions and schemes which were vastly ambitious.

* * *

The Parabag had excited military interest the during the war, and finally testing began at Wright Field in Ohio. George's original idea was to jump twelve paratroopers together in a kind of enclosed bag suspended from a huge parachute, to be immediately followed by the Parabag itself which held sufficient supplies for an extended mission. Nothing came of the testing at the time, though years later, and with no credit whatever assigned to George Hutchinson, a form of the Parabag was indeed used for supply drops to American troops.

When one scheme floundered, George took up another. He wrote up an innovative plan designed to help airlines avert hijackings by taking a few simple precautions. Since there was no War on Terrorism to incite alarm at the time, his proposals died on the vine. The airlines weren't interested.

A ground cover device for baseball diamonds seemed promising. Designed for use during rain, the only problem was that of torque. It unrolled unevenly. Too impatient to go through the patent application process, George abandoned the idea. Years later, the torque problem solved, the cover idea was successfully developed. Once again, George just missed the boat.

* * *

But, George's most intense frustration by far, lay in the lack of interest in his grand plan for "Bloodless Warfare." This scheme involved parachuting great numbers of troops into an enemy city under the radar. A chemical would be employed to temporarily paralyze everyone except the landing troops. All buildings, utilities, and defensive and offensive capabilities would be neutralized while an elite corps of highly trained commandos took over. This peacekeeping force would thus constrain destruction of life and property, yet effectively neutralize rogue governments.

As a first step, George wanted to make a film to demonstrate his plan visually. He turned to the heads of major studios and to entertainment moguls such as CBS for help in getting his movie produced. When it wasn't forthcoming, his letters grew ever more strident and insistent. He wrote to Darryl Zanuck in 1968:

In the twilight of your life, how would you like to produce a motion picture that would earn for you the gratitude of all humankind...a motion picture that would show in the most minute detail...a new concept of high speed force...that can halt any type of war-like aggression within 72 hours!!!!

And to the head of 20th Century Fox:

Surely you are a man of character, who could not refuse to evaluate a formula that offered so many opportunities to earn the respect and gratitude of all humankind.

Next he turned to politicians. George was highly insulted that President Nixon wouldn't meet with him to listen to his proposal. He turned instead to former President Lyndon Johnson, possibly because of the latter's friendship with filmmaker Jack Valenti. In this letter he stressed that he, George, was no stranger to hobnobbing with Presidents:

In the White House, in March 1939, President Franklin D. Roosevelt personally told me that he had learned more from me in one hour than he had learned from any other man in his life during an equal period of time.

As his plan was ignored or turned down by various leaders, George's tone of moral indignation grew and probably worked against his own purposes. From 1968 through 1980, he refused to abandon his vision. To him, the indifference to this most idealistic of his visions was incomprehensible, and deeply hurt his pride.

One of the final enterprises as a family undertaken by the Flying Hutchinsons was the television game show *World Tours*. Appearing on a local Baltimore station, *Tours* attempted to stump guests by asking them geography questions (formulated, naturally, by George) which they had to answer before a model airplane reached the map destination on a wall. Janet had her friends appear as contestants, sneaking them the answers in advance.

* * *

While his own enterprises failed to bear much fruit, George Hutchinson watched Janet's little nursery school grow and grow. When the student body numbered twenty children, he made his move.

One day, out of the blue, he sat his daughter down for a serious talk.

"I think the best thing for the future of the school is for me to take over as head at this point. You've gotten things off to a good start, but we could do much better. I'll handle the management and publicity angles, and you can keep on teaching. I've got some really great ideas...The way I see it, the sky's the limit here!" As always, George's tone was totally assured and forceful.

Janet stared up at him, literally speechless. Since babyhood she had learned to follow this man and to defer to him in all things. Openly challenging his decisions had never been an option. So while uncertain and rebellious emotions churned in her breast, the face that Janet turned to her father was impassive.

"Well, sure, Pop, if you think that's what's best." The words seemed to stick in her throat.

Years later, George commented in a newspaper interview about the startup of Kiddie Kollege.

"I did it for my daughter...her marriage had broken up and she had a three-year-old boy. We always tried to do things as a family."

Fifty-odd years later, Janet sighed as she looked back on the ease with which she had relinquished control of the business she had built into her father's hands.

"If it were today, it's be a different story!" she said forcefully. "But if I'd resisted him, the school would never have become what it did. He was a promoter; I wasn't. He really did build that school up in ways I probably wouldn't have. So who's to judge?"

* * *

The next few years would be full of growth as the popularity of Kiddie Kollege spread like wildfire through the upwardly-mobile Ruxton area. There was a waiting list to get in. Expectant parents signed their children up before they were born. Offspring of prestigious old Maryland families, including the Eisenhower grandchildren, filled the school to overflowing, necessitating three building extensions. The Hutchinsons ended up purchasing the house next door, owned by George's best friend's widow, to provide more classrooms as the enrollment climbed to a high of around 100 children.

Eventually, a spacious apartment was built over the school and Janet and Robbie moved out of her parents' home and into their own digs. Kiddie Kollege became a true family enterprise. Both Blanche and Kathryn taught there for many years.

Articles about Kiddie Kollege activities, graduations and awards appeared every week, it seemed, in local newspapers. Usually "Pop George", as the children called him, took center stage in the photos, his shock of white hair easily recognizable. Janet, the real power behind the success, appeared less often.

George had a model railroad train and track built around the property. The kids could play and sit in it (especially for photo ops) but it didn't move. He loved to boast to the public that "We have children here from 21 different countries!" Actually, they were never all there at one time, but Kiddie Kollege did have an international flavor. Foreign doctors sent to train at Johns Hopkins usually sent their children to the school.

As long as George confined himself to promotion, things went well, but, characteristically, he liked to stick his fingers into every pie. When he fiddled around with the curriculum, where he didn't have expertise, Janet fumed in silent frustration. George had little tolerance for any ideas but his own, and his daughter often had to work "around" him to put her own very solid plans into practice.

Fortunately, she had wonderful help. In addition to her good right arm Bettemae, who would remain with her at the school for the next 29 years, Janet's other mainstay was Doris, an African-American assistant and teacher. She was the absolute favorite of all the children.

"Doris," recalled Janet, "only had one expression. A big smile."

Janet was also frustrated when her father interfered in the hiring. He tended to pick the most attractive and well connected young women for positions instead of the most qualified applicants. George also had a habit of sneaking up on attractive women and giving them a pinch on the behind, something he had always done, teasingly, with Blanche and the girls. The women were highly annoyed, but in the days before sexual harrassment suits, they simply learned to ignore it.

25

HORSES

Robbie, growing bigger by the day, shared his mother's enthusiasm for horses. When he was about eight years old Janet sent him to Mrs. Bosley's Riding Academy to learn to ride. Walking and trotting came easily for the boy, but he really wanted to canter.

He wanted to, but he was scared to death of it at the same time. The first time he tried to urge his mount into a canter the horse just wouldn't cooperate, and Robbie became so frustrated he cried. His mother saw. Janet stalked into the ring, grabbed the reins to Robbie's horse, and led them out of the ring. She didn't say a word to her humiliated son.

"Can I borrow a horse?" she demanded of the startled instructor. Not waiting for a reply, Janet grabbed a bullwhip down off the wall. "To motivate the horse," she tossed over her shoulder, marching toward the paddock. Wordlessly, she saddled up a new mount and turned to her son.

"Get on."

Hesitation didn't seem to be an option. The horse knew Janet meant business, and so did Robbie. In a few minutes, the boy was cantering around the ring with the cool wind drying the hot tracks of the tears on his cheeks. Joy thrummed in his blood. From that point on, riding became both fun and a passion for the young man, as it had been for his mother.

Robert Simpson, a tall, tanned Californian with a successful international business paused to look back, decades later, on what he had learned from his mother that day.

"I learned that when you're afraid, afraid of anything, you deal with it by going ahead right through it. You don't let it stop you," he recalled.

* * *

With her father so heavily involved in its operation, Kiddie Kollege began to seem more confining and less stimulating to Janet. Tired of the same old grind, she longed to turn once again to something she had a passion for.

Janet approached her father with her mind made up. "Daddy, I want to leave the school for awhile. Sort of a sabbatical. I want to get my license and train racehorses."

George was not pleased. He loved basking in the limelight as director of the Kollege, but had always left the running of everyday operations, the nitty-gritty hard work, up to Janet. This time, the harder George tried to talk his headstrong daughter out of leaving, the more firmly set her intent became. George didn't think her latest enthusiasm would last long, but finally he was talked into supporting her decision. He even called upon his old friend, the racing commissioner, with his determined daughter to get her license application rolling.

Janet was a natural as a trainer from the start. Her deep love and instinctive understanding for horses combined with ironclad determination formed a potent blueprint for success. Once Janet set on a course, she gave herself to it a hundred percent.

She started with the horse her beloved Aunt Kathryn had given her. Fancy Fleet, son of the famous racehorse Count Fleet, helped Janet really establish her credentials in her new career.

Auntie Kathryn had given Janet Laudy, her first beloved horse, and was perhaps more simpatico with her niece than anyone else in the family. Like Janet she was active and athletic, an excellent golfer. She loved bowling. When she and Janet visited the lanes, Auntie insisted on staying until closing time and then shooting craps afterwards with the pin boys. She got right down on her hands and knees with them.

Fancy Fleet was "mean as anything", but did increasingly well at the track, and Janet began to earn her reputation as a trainer. Soon, the petite blonde was a familiar sight at the Maryland tracks: Pimlico, Hagerstown, Timonium.

Janet believed in establishing a bond, physical and emotional, with each horse she worked with. She mucked out their stalls, groomed them, whispered in their long, silky ears. She even warmed them up for races by running them around the track.

Before long, the small blonde with the special way with horses was being sought out by owners. She ended up with a roster of ten horses. Soon she was traveling out of state to the Shenandoah and Charlestown tracks in West Virginia, pulling horse trailers behind.

A couple of painful falls convinced Janet to forgo doing her own warming up. Robbie, already a very popular exercise boy, stepped into the breach.

Robbie got up hours before his first period classes to perform his duties. He earned three dollars for every mount exercised and soon was bringing home $40 to $60 a day, and more on weekends.

Fancy Fleet in particular was a challenge. The horse was so wild and strong he had to be galloped the *wrong* direction around the track. Otherwise, he got carried away and was too difficult to control.

* * *

During these years Janet carried on an affair with a fellow trainer. Although they shared the life of the track, she never really considered him a boyfriend.

Janet bought Brasher Falls, an unbroken six-year-old, cheap. Brasher was docile and easy to train, and in their intense traning relationship the two developed the kind of love and bond Janet had only shared once before, with Governor the lion. She even slept with Brasher in his stall three nights a week, simply because she wanted to be with him.

Like the famous underdog Seabiscuit, to all appearances Brasher Falls was a long shot. He was lethargic in the paddock and slow out of the starting gate. But Janet soon discovered that he could, and would, run like the wind when he was given his head at just the right moment.

Remembering, Janet's eyes grew dreamy with time. "At the mile and a sixteenth run, you'd have to just let him go when he got to the half mile pole. Show him the whip and go *click click* and hold on tight." More than once, it happened that Brasher came from at least twenty lengths behind to win by more than five lengths.

Devastatingly, Janet lost her horse when he was claimed right out from under her during a race. She was stunned by the loss. It felt like she'd been forced to give up a child.

Brasher Falls never won another race. He simply refused to run for anyone else.

* * *

Janet's life was centered at the track from 1968 through 1978. Although she trained horses for ten years, she was only away from her Kiddie Kollege duties for seven. For three of those years, she split her time between the school and the track, because of her aging parents' increasing need for her presence back home. These years segued into a nearly intolerable period of overwork that set her up for burnout.

Each day Janet picked up and delivered 22 children to and from school. On the track scene, Janet was transporting racehorses to racetracks up to two hours away as well as maintaining her normal training duties.

Once, on her way to the Shenandoah track, Janet glanced out her side window and was shocked to see her own horse trailer rolling *beside* her down the opposite highway lane. It hadn't been hitched up properly and had come loose. Fortunately the trailer veered into the grass by the side of the road and bounced to a stop, still upright. The horse was unhurt, but he didn't win that night.

Janet sometimes functioned on as little as two hours' sleep a night during this period. When she fainted at the track one day, it seemed like a wakeup call. She couldn't continue this pace.

Sheer exhaustion, plus the uncertainty of the horse business, contributed to Janet's decision to return full time to the school. But the main motivator was how badly she was needed back at the demanding family business Kiddie Kollege had become. She needed to take over more of George's administrative duties and public appearances. Blanche, too, was beginning to have problems. In the beginning stages of a protracted and gradual slide into Alzheimer's, she had increasing difficulty in keeping the schedule straight and remembering the names of students. Kathryn helped her cover up, but Blanche herself was aware of the problems, and ended up voluntarily leaving her teaching duties.

The Hutchinsons had always functioned as a team, and the pull back into the Kiddie Kollege orbit was a strong one. Still, giving up the track was a true loss and Janet felt it deeply. She bought one more racehorse and had a friend train him, just to keep her hand in.

* * *

Like his mother, Robert Simpson was possessed by a love of horses from a young age. Like Janet, too, he set high goals for himself and drove himself mercilessly in their pursuit. Through his teen years his involvement at the racetrack continued, even after Janet gave it up.

Socially, though, he suffered. Before and after school, instead of hanging out with the other kids, he was usually at the track. Robbie matured early, his work with the horses giving him a sense of independence. But as he entered adolescence, Robbie's relationship with his grandfather grew more difficult. George had difficulty sharing the stage with another, younger male who, like George himself, had an independent streak a mile wide. Robbie was entering the normal rebellious, testing stage of the teen years. This added up to a recipe for disaster for two such strong and volatile personalities.

The inevitable confrontation occurred while Janet was still training horses at Dover, mainly living at the track. Robbie was often alone in the apartment he shared with his mother. As he had done with his girls, George tried to impose strict rules and curfews on his grandson. He would stay up to check on when the young man made it home.

Late one night, Robbie was caught sneaking back in. His grandfather glowered at him from the top of the stairs.

"Just where have you been?" George barked. "Do you have any idea what time it is? I'm not having this!"

Robbie shrugged, feigning indifference, and continued on up the stairs. "I've been to see my girlfriend, what do you think," he mumbled.

"What did you say to me?"

"It's none of your damn business where I've been!"

George lunged at his grandson as he tried to push past him. He kicked at him and Robbie stumbled back down the stairs. He charged right back up them again.

Again George kicked out and back down Robbie tumbled. He picked himself up, seething but silent, and kept coming. One more time he was knocked down, one more time he remounted the stairs.

"You stop it!" George yelled as he lunged at Robbie once more.

This time, the younger man blocked his grandfather's blow, and instead rode him down the stairs on his back as George fell. Robbie was clutching at his grandfather and sobbing, but he held on.

George was not really hurt physically, but more had been bruised than his bones during this encounter. He was beside himself with outrage. Holding Robbie at bay like an apprehended criminal, he called his brother Leonard.

"Call the police! Robbie jumped me in the dark and threw me down the stairs. He's gone crazy. No, there was no reason for it at all. Just call the police!" The older man was apoplectic with shock and rage.

He glared at Robbie. "You're getting out of here tonight. You're not living under my roof for one more minute!"

Shaking, Robbie called Kathryn's son Dwight.

"Just stay put, Rob. I'm coming to get you," his cousin said.

Robbie spent that night at his Aunt Kathryn's. In a sense, he left home for good that night. He would never live there again as a child.

George, of course, immediately notified Janet.

* * *

Robbie was sorry later for his part in the fiasco, but no one really listened to his side of things. Much as she loved her son, Janet didn't feel as if she had a choice in the matter. How could she disbelieve her father, the authority she had lived under all her life? That original bond could not be breeched.

Janet felt torn, confused, and angry at Robbie for "bringing this on." Mother and son did not really communicate for the next two years. Inside, Janet steeled herself to silence and suppress her troubled emotions.

As for Robbie, he could not comprehend or accept that his own mother wouldn't believe him. Bitterly, he turned aside from the home that seemed to no longer hold a place for him. From now on, he would show his family that he didn't need anybody. At the age of fifteen, he was on his own.

The teenager went to live at the only other home he knew - the racetrack. He worked full time as an exercise boy, dropping out of high school for a while. Later he put himself through night school in Hagerstown. He was too young to drive, so out of his earnings he paid hangers-on at the tracks to give him rides to and from school. Later, he purchased his own motorcycle.

It was to take another catastrophe to bring mother and son back together again. Robbie's first accident came when he went over the handlebars of his cycle and badly hurt his knees. One knee required six surgeries over the next few years to repair the damage.

Janet visited Robbie in the hospital. Her heart went out to him as she saw the pain he was in. Still, a gap remained between them. Too many things went unsaid.

Because of his injury, Robbie could no longer gallop the horses and became a gate handler instead. While schooling a horse out of the gate at Pimlico, the horse shied and crushed Robbie's leg against the rails. This was major damage. During his recovery, Robbie dropped from 128 to 105 pounds. Prior to his accident, Robbie had been planning to become a jockey. Now those plans, like his leg, appeared crushed.

Janet rushed to her son's side once more. He truly needed her, and this time her heart opened fully to him. The two would never be parted emotionally again. They forged a mother-son bond of unusual strength and depth.

* * *

Robbie continued to develop along his own course. Just two years after the last accident, he was managing a whole string of horses for an owner down in Florida. Times were good, then not so good. When that job ended, he went to walking horses for $65 a week for a major Florida trainer. Once, he went door to door as a house painter. He became an assistant trainer and eventually ended up with his own string of horses. At that point he felt he had done it all in the racing field, and looked around for a new challenge.

26

WINDING DOWN

When Janet decided to give up the race track once and for all, life felt strangely flavorless. Things had not seemed the same to her, true, since the forced sale of Brasher Falls. As with the loss of Governor the lion many years prior, she would never forget the friend of her heart. Still, she had more than enough to occupy her as she settled into a routine of work and intense development at the school.

Perhaps the school was so successful because it was so much fun for the children. Christmas pageants and the Kiddie Kollege Circus, with exotic costumes and live animals, were more like Hollywood productions than preschool doings. At one performance Janet sang a solo of "Over the Rainbow" with all the children joining in at the end. Parents were thrilled.

Animals had always played an important role in Janet's life, and she continued to own dogs. Muffin, a Jack Russell terrier, was with her for many years and was of course a great favorite with the children. Muffin took center stage during circuses and pageants. Poor Muffin was regularly costumed in a gingham dress and sunbonnet and pushed around in a doll carriage. When Janet sang, Muffin loved to howl along in accompaniment.

Muffin had one litter of puppies. One tiny female was apparently born dead. Bettemae, calling on her nursing skills, picked the tiny creature up by her hind legs and flicked it with her fingers. The pup started breathing with a gasp. The rest of the litter was given away, but little Buffy stayed.

After Muffin's death Buffy replaced her mother on the Kiddie Kollege showbiz stage. When the gingham dress came out, though, Buffy was smart enough to make herself scarce.

When the potbellied pig craze hit, Janet just couldn't resist. The creatures were so ugly, yet so cute. She bought Beauregard (Bo for short) for $1200. The children absolutely loved him. Bo was smart, full of personality, and could run like the wind.

The problem was, Muffin didn't deal well with the challenge of sharing space and attention with a pig. The little terrier began to act out her stress by throwing up her food and howling.

One day, Janet had to go to the store and put Beauregard in the bathroom so as not to leave the two animals alone together. When she returned she found the shower curtain shredded, the tile on the floor gnawed, and even the folding wooden bathroom door chewed up.

Soon thereafter Muffin, in the middle of the night, appeared to have a complete nervous breakdown. She wouldn't stop howling and crying piteously. She threw up constantly. Still in her bathrobe, Janet rushed her to the emergency vet.

The professional prognosis: "This dog is stressed out. You must choose between the dog and the pig."

It wasn't really a contest. Janet loved Bo, but the dog came first. Still, she hated selling the pig so much she hopped on a plane to go visit Robbie, leaving the selling of Bo up to others while she was gone.

* * *

Kiddie Kollege continued to thrive. Graduation ceremonies grew so large they eventually had to be held at the local college. Janet became increasingly overworked and overwhelmed. What once had been a smoothly-functioning family enterprise was beginning to feel like more and more of a burden.

As George and Blanche aged, things were changing. Their close and loving bond, however, never wavered. This was brought home to Janet vividly in an incident during her parents' seventies. Passing through the hall outside their bedroom, she knocked, received no answer, then pushed the door open. George and Blanche, caught in the act as it were, looked up at her from the bed and just smiled.

Janet was far more embarassed by the encounter than the aging sweethearts, who seemed to take it in stride.

Janet's parents' involvement in the active running of the school was winding down in the late Seventies. And as George was needed less at the school, he spent more of his time gambling.

George had always loved the track, but now it became a fulltime occupation. He was on the phone seven days a week placing bets at tracks all over the country. When George lost, he tried to cover it up, just like an alcoholic hides his drinking. One of the little girls enrolled in the Kollege was the daughter of George's local bookie. He was terribly upset when the man died owing him money. Still, as he aged, George became more mellow and less controlling of those around him.

Conversely, Blanche's creeping Alzheimer's left the formerly sweet and serene wife and mother prey to vicious mood swings. For the first time in her life she burst out in anger at those closest to her, even George. Little things irritated and confused her, and she became frightened by her own loss of self-control. An argument with her husband exploded one afternoon and she chased George across the lawn with a rake, in front of the students. Paranoia bloomed as a side effect of the vicious disease. Surprised by Kathryn's youngest daughter in the kitchen, Blanche grabbed a butcher knife and turned on the startled girl, mistaking her for an intruder.

Tragically, the same thing was happening to Janet's Auntie Kathryn down in Ft. Lauderdale. Janet had visited her every year for two weeks in the summer. The last visit found her aunt disoriented and so confused she didn't know where her own bathroom was. Janet flew her back to Maryland. On the flight her aunt thought she was on a train and kept trying to run for the door. Soon she was ensconced in a nearby nursing home.

Social life at the Hutchinson enclave, which had once been so active, slowed almost to a stop. Sunday afternoon get-togethers on the lawn, which in the past would have seen up to 20 people picnicking, playing badminton and volleyball, tossing horseshoes and "just visiting" now dwindled and died. The weekly poker game with her father, uncle, and their friends, which Janet had always joined in on as the "token female," ceased.

In the late eighties, George's health began to fail seriously following a prostate operation. The once vibrant, dynamic patriarch of the Hutchinson family was fading fast. He became confined to bed and never really recovered. The day came when Robbie had to carry him downstairs to be transported to the hospital and finally to a convalescent home. George Hutchinson died of heart failure in 1989 at the age of 87.

Blanche floundered. The vital center of her life was gone, and it seemed that George had departed taking her heart and her will to live with him

* * *

Janet knew that should she continue on with Kiddie Kollege, she herself would eventually break down. Her decision, when it came, was fast and sure. Kathryn, now widowed by her second husband, had moved into a new house in Ponte Vedra Beach, Florida. Janet had spent Christmas nearby at a luxurious spa on the ocean. From her balcony she gazed out over the rolling dunes and white sand to the sparkling Atlantic. She inhaled a deep lungful of fragrant sea air. *I could live here,* she thought.

Fortuitously, the lot right next door to Kathryn's home was empty. Janet marched over to the real estate office that very day and ordered a house built on it.

When Janet returned to Maryland she wasted no time. A former Kiddie Kollege student bought the property for "peanuts," later converting the facility into a beautiful home that he turned over for a tidy profit. But, Janet didn't look back. She was Florida bound.

Blanche could not be left behind. For a time after the move, Blanche alternated staying with Janet and Kathryn, but continued to decline rapidly to the point where she didn't even recognize her daughters. Finally they simply couldn't care for her, and after serious scouting found a nursing home that would provide excellent care for their mother in her last days. Blanche died in 1993.

* * *

Surprisingly after so many years of nose-to-the-grindstone, Janet found she didn't miss the school or the world she left behind. Florida was a new start and a new adventure. She loved the sun, the water, the beach and her new leisure.

Janet wasn't consciously lonely. "I loved living alone. I really didn't understand what loneliness was," she recalled.

When she moved, Janet still had her little terrier Buffy, daughter of Muffin. Soon after the move Buffy developed awful seizures and began bleeding from the rectum and mouth. A visit in the middle of the night to the emergency veterinary clinic was futile. When Janet came back the next morning, she was told that Buffy had died of a heart attack.

It was unthinkable for Janet to be without a pet for long. She planned on getting another dog, but Hillary, Kathryn's youngest daughter, showed up at her door one day with a basket of six-week-old kittens. Janet picked out two to keep. Kathryn followed suit.

Janet named the female kitten Buffy, after her beloved Jack Russell, and her brother Buddy. The beautiful half-Persians became an inseparable part of Janet's life from that point on. Buddy grew into a gentle black giant with enormous golden eyes. Buffy had the elegant markings of a Himalayan, with a coat so dense that Janet had her clipped in the summer to resemble a miniature lion, with only her ruff, paws, and tail tip left long. The two were enormously affectionate, and loved to lie curled in each other's arms on "their" footstool for much of the day. Janet, with her usual skill with animals (even cats) trained them NOT to venture onto the other furniture, and to retire to their own crates on the screened lanai precisely at bedtime. Janet swore they knew and responded to the commands "eating time," "bedtime," "soon" and "in a minute."

One morning, standing in her kitchen, Janet found herself thinking about her mother. Blanche's death had come so swiftly, and it had bothered Janet that she hadn't been able to cry at the funeral. Somehow she had remained frozen and detatched from her own grief until this day, when out of nowhere the tears poured out of her with an intensity that left her breathless.

Such territory had never been Janet's home turf. Although a woman who felt deeply, she had lived much of her life out of touch with her own emotions and confused by them. She was much more comfortable in the world of action.

All that was about to change.

27

DARKENING SKIES

Alcohol gave me wings to fly
and then it took away the sky.
Anonymous

    Janet had no real issues with alcohol prior to her retirement and move to Florida. Drinking had always meant socializing with friends. Other than a few occasions when she ended up having too many cocktails at a party, alcohol hadn't yet caused any problems in her life. Moreover, she knew almost nothing about alcoholism as a disease. Janet wasn't aware of ever having been acquainted with a "real" alcoholic.
    Still, Janet worried about her old friend Doris, who still called every week. Apparently Doris had developed the habit of drinking alone as she grew older. Her voice over the phone sounded bitter and depressed, but when Janet asked if she could help in some way, Doris brushed it aside angrily.
    In total ignorance of what addiction was all about, blind to the horrors that lay ahead, Janet's own drinking patterns began to change as she settled into her new Florida lifestyle.
    Retirement may have freed Janet of responsibilities and burdens, but it also challenged her to develop a new community she could feel a part of. Janet had always been intimately involved with family and with work. As part of the Flying Family, as a young WASP, as a horse trainer and owner of Kiddie Kollege, she had always felt needed and "a part of things." Now her parents were gone, her son was in California, there were no work demands, and

for the first time ever she was not part of a close knit group. What was Janet to feel a part of now? Where were "her" people?

It happens to many unsuspecting new retirees and seniors. Businessmen and women who have been so busy and involved most of their adult lives in their careers that drinking hasn't been a noticeable factor may find that alcohol takes on a whole new presence after retirement. The same emptiness can take hold for widows or widowers. Even when retirees have looked forward with anticipation to their new leisure, the very lack of structure can lead instead to a sense of emptiness or anxiety.

Like a worn-out suit of clothes, identities that came along with the job title or the old familiar role are set aside. Questions such as *Who am I?* and *Do I matter?* may arise for the first time. Too many find that another drink quiets the discomfort.

* * *

It is said that Nature abhors a vacuum, and soon Janet's need for connection led her to seek for companionship in ways that were to start her down the dead-end road of alcohol abuse.

In her usual friendly fashion, Janet had made friends with a close neighbor, "Maxine". Maxine, years younger, was a daily drinker. She began showing up at Janet's door, wine glass in hand, at all hours. Before long it seemed natural to join right in, and a habit was established. Afternoons meant drinks and conversation with Maxine on the lanai. Maxine could put away a fifth and a half of wine in an afternoon and Janet's own drinking escalated to keep pace. She had no idea that she was beginning a descent into a kind of hell that would take away her power of choice over her drinking and her life.

Maxine introduced Janet to her favorite bar in a nearby upscale Ponte Vedra Beach restaurant. Janet volunteered at the Humane Society on Thursday mornings, taking puppies around to local nursing homes to cheer the residents. She began stopping by the lounge after she finished her rounds on Thursdays. Soon, more days than not, she could be found among the regulars that hung around the bar from about 1:00 PM on, sipping and chatting throughout the long sunstruck afternoons.

Maxine's drinking had progressed to the point where she preferred drinking at home and was nervous going out. Usually

Janet went alone to the bar, but drinking at home with Maxine was a part of her everyday routine as well.

Often they were joined by Maxine's boyfriend "Sam", and the three would sit around and talk for hours about "everything under the sun." Janet enjoyed being part of this fellowship, alcohol-based though it was. She felt like a part of things again, and needed, as Maxine turned to her more and more for advice and comfort following her boozy fights with Sam.

Several times Janet went along with Maxine on Sam's boat. Maxine was driving, and speeding, and of course drinking, when they were pulled over by a Marine Patrol boat. Fortunately or not, they got by that time with a warning.

Another time the fight next door raged out of control. Maxine was beside herself, throwing plates, screaming, assaulting Sam. The police were called, and Maxine was arrested. She spent the night in jail. The next morning she wept over the phone to Janet, begging her for bail money.

Some of Maxine's actions seemed completely irrational even to Janet. She would crawl out the bathroom window of her own house late at night, leaving Sam asleep, and come knocking on Janet's door. Another night, Maxine got mad during an evening's partying and marched home, leaving Sam behind. Sam pulled his chair up close to Janet's and reached for her. He kissed her. It might have gone further, but "I wasn't drunk enough," Janet recalled. Sam had the opposite problem. Tiny Janet had to practically drag the six-foot tall man home. He fell twice on the way.

Dimly Janet realized that Maxine was veering more and more out of control, but the seductive intimacy of being needed, being part of what felt a little like family (albeit a dysfunctional one) was too strong. It was even harder to see her own developing dependence on alcohol. It no longer seemed strange to have an eyeopener in the morning. When she started drinking alone as well as in company, it was a sure sign that social drinking was a thing of the past. The first stages of the disease of alcoholism had taken hold with a vengeance.

* * *

Janet was far from alone in her developing problem. Americans over 65 constitute the fastest-growing population of alcoholics in this country. The reasons why reflect the changes in society and in the population of retirees and seniors as a group. Americans are living longer, and increasingly adopt an active lifestyle in retirement that can include lots of social drinking. We are a self-medicating society. Many seniors learn to cope with loneliness, depression, isolation, and failing health by turning to a substance.

But tolerance diminishes as the body ages, and the cumulative effects of alcohol abuse affects senior health dramatically. Doctors are slow to address this issue with patients for whom the word "alcoholic" is still a matter of shame and secrecy. Thus most senior alcoholics remain undiagnosed and untreated.

Adult children may be "worried sick" about Mom or Dad, but because there is still so much outright ignorance of the fact that it is a disease and not a moral issue, it can be hard to get past the barriers of self-protective denial to the truth. This is especially true of that post-WWII generation of seniors, self-reliant and accustomed to "sucking it up" instead of airing their dirty linen in public, to which Janet belonged.

* * *

Meanwhile, back at the bar, Janet's habit of stopping by on Thursdays expanded to every afternoon. Her bar friends grew to count on the petite blond to show up most afternoons at 1:00 PM, rain or shine. She became pals not only with the bartender but with two younger women, both very attractive, who found in Janet a willing and sympathetic listener. Through a blur of booze, their ongoing laments of crises and man problems became more entertaining than a television soap opera.

"Shane" was a forceful personality who easily swept Janet into her dramatic orbit. "Anita", her bosom buddy, waltzed into the bar one day pushing a baby carriage. Her vodka bottle was stashed under the blanket. An attractive divorcee with four children, Anita competed fiercely with Shane for the male attention that seemed to be their main preoccupation. The duo loved to pursue older, well-to-do businessmen and bragged about the money and gifts lavished upon them by grateful admirers.

Janet was shocked by tales of them both sleeping with the same rich older gentleman. Although the two had been fast friends since their early teens, they criticized each other to Janet incessantly.

Afternoons at the bar followed by dinners at a local seafood restaurant became part of the drinking lifestyle that was fast becoming Janet's world. The older woman always seemed to get stuck with the tab.

* * *

Kathryn, living just next door and increasingly concerned, tried to convince her sister that she was just being used. But to Janet, it was intoxicating in more than one sense to be hanging out with the younger gals, to be their confidante. It was a bit like being back in the WASP barracks after lights out. These girls seemed to need her, and Janet had always been fiercely loyal to her friends.

Her own growing consumption of alcohol (wine, mostly) made everything recede into a boozy haze. Blessing or curse, she never suffered from hangovers, thus being spared the pain that functions as a tormenting wake-up call to many problem drinkers.

Janet's progression from social drinking to full-blown alcoholism did not follow the typical pattern in that she began abusing alcohol late in life, and once she did, the downward spiral accelerated fast. But then she had never been one to do things by halves.

By 1998, Janet was well into the phase of active alcoholism that can lead to the phenomena known as "hitting bottom." Hitting bottom signifies the point at which the life situation of the alcoholic becomes so painful or catastrophic that she can no longer deny she has a problem. For many, hitting bottom comes through a major loss. A relationship, a job, one's health, anything of value can be claimed by the rapacious creditor alcohol. Most of all, and particularly for the female alcoholic who is affected by the shame of not being able to "drink like a lady," hitting bottom means a loss of self-esteem. She can no longer live with the person who stares back from the mirror every morning. For Janet, hitting bottom took the form of a disastrous car crash that almost killed her.

* * *

The evening began undramatically. A friend from the bar, an older woman, had invited the four "girls" (Janet, Shane, Anita, and the bartender) to dinner at her beautiful Beaches home. Janet played chauffeur. The food was wonderful and the liquor flowed. Everyone got sillier and laughed louder as the evening wore down.

Still, Janet knew she had to drive home and so kept some degree of control over her drinking.

At one point Janet noticed Shane weave her way to the bathroom, where she suspected she was adding some pharmaceuticals to the supply of alcohol already in her system.

It had already started raining when everyone finally said their goodbyes and headed for home. Janet dropped Anita off first, then the bartender at her trailer. When Janet returned to her car, Shane had planted herself firmly behind the wheel.

"What do you think you're doing? You're too drunk to drive!" No way did she want to ride with Shane driving.

Shane just laughed. "Oh, come on and get in! We're both drunk. You're getting soaked!"

Vainly Janet tugged on her arm. "It's my car!"

Shane shook off her hand impatiently. "Yeah, well I'm still driving. I see a hell of a lot better than you do in the rain."

Shane was a lot bigger than Janet, and there was no reasoning with her. With a sinking feeling in the pit of her stomach Janet got in the passenger side and they took off, wheels fishtailing on the rain-slick pavement. In her anxiety Janet neglected to fasten her seatbelt.

The road before them was a rain-slicked ribbon slicing through the darkness. Twenty minutes later, taking a curve too fast, the tires lost purchase on the slick pavement. Shane skidded, desperately fighting to regain control of the wheel. Janet looked up just in time to see a big tree rushing right at her side of the car. She had time to register Shane's piercing screams, the crashing, crunching sound of impact as the world exploded. Janet's head went through the windshield, and then all was silence and darkness for a long, long while.

* * *

Shane escaped with just a scratch over her left eye. When Janet went to the dump to view her totaled car months later, she saw that the passenger side was accordion-pleated and crumpled while the driver's side remained untouched, as though the car had been cleaved neatly down the middle with a knife. The windshield was shattered and a large hole marked where her skull had impacted with the glass. She had been lucky to emerge alive.

Out in Carlsbad, California Robbie was stunned when he picked up the phone and heard the Doctor's words. *There's been an accident...prognosis for your mother doesn't look good...possible brain damage.*

Living across the country, Robbie had had no idea of his mother's developing problems with alcohol or of what her life had become. He dropped everything and flew to her side.

When he walked into the intensive care room a day later and saw the tiny, white-draped form beneath the sheet, he couldn't believe it was his fireball of a mother lying there so ominously still.

Janet had broken both shoulders and elbows. Her skull had splintered the windshield and the doctors didn't know yet what kind of brain damage might result. A broken rib had punctured a lung. A deep gash from the broken glass sliced across one cheek like a dueling scar.

"Don't get your hopes up too much," the doctor warned him gently. "She's still deeply unconscious, and considering the extent of her injuries...well, at her age, there may be very little hope."

The doctor, however, had reckoned without Janet's ironclad constitution and force of will. She struggled back, although recovery was slow, particularly from the shoulder injuries. She had pins in both shoulders and screws in her elbows. Her shoulder would never return to full function, and over the months she would undergo six operations to repair the damage.

Janet remembered opening her eyes and seeing her son's worried face so close to her own. She was on morphine for the pain, and green dragons were climbing the walls of her room.

"Mom, you're hallucinating!" Robbie insisted.

Janet went from University Hospital in Jacksonville to Shands to Arbor Place for convalescent care. She was hospitalized for a total of three months.

28

TOWARD THE LIGHT

Janet's injuries were painful, but more painful still were the emotional scars. She was filled with shame and rage. Inwardly she seethed with resentment at Shane, who she blamed for putting her in this hospital bed while she escaped unscathed from the accident. But when she finally looked in the mirror one day, she realized that forgiving *herself* would be even more difficult.

Janet had reached an emotional bottom. Looking at the woman in the mirror, she realized she could live with herself no longer. The face that stared back at her seemed ugly and twisted and she hated what she saw. Shame burned in her heart. Yet somewhere deep within, a light came on.

I can't do this anymore, she said to herself. *I'm through.*

* * *

Janet's first steps toward a new life without alcohol came while she was still recuperating at Arbor Place. She attended her first self-help meeting for alcoholism.

At first she couldn't believe that these well-dressed, smiling people could have anything in common with her. They were all claiming to be alcoholics! She found it diffucult to raise her head, to meet their eyes. Yet even at that first, confusing meeting, as she listened to the stories of out-of-control drinking that rivaled or surpassed her own, she had a feeling she was among people who truly understood. Women reached out to her after the meeting. Several would become close friends in the months to come, members of the same meetings Janet attended near her home on the beach.

Although many newly recovering alcoholics must struggle for a time with denial, Janet was thankfully well past that point. She had no trouble admitting she was an alcoholic. The shame and guilt she felt about that fact, however, was a bigger problem. It kept her, for a time, from opening up and sharing much in meetings.

It was hard for Janet to get past the initial feeling that she had failed. After all, she had been able to meet every real goal she'd ever set for herself in her life. She had tremendous drive, self-reliance, and willpower. Why had it failed her in the end?

* * *

Non-alcoholics have a misconception that most alcoholics are weak-willed. The opposite is usually the case. Living in active alcoholism isn't for sissies. Many alcoholics suck it up and work day after day with with hangovers, the shakes, the demoralization that comes with the territory. It's hard work. The problem isn't lack of willpower but the fact that a disease doesn't *respond* to willpower. Just try telling a diabetic to use his willpower and throw away his insulin. It simply doesn't work.

Janet would learn over the next few months that once alcoholic drinking had become a habit, willpower was doomed to fail her. The illusion that the alcoholic will ever be able to successfully drink like "normal people" by exerting still more willpower and control must be defeated. Usually, as with Janet, it takes some fairly demoralizing and painful life experiences to reach that point of surrender, where willingness rather than willfulness leads on to recovery.

Janet had been raised to be strong, independent, and keep her feelings pretty much to herself. In recovery she learned that opening up to the group was necessary and that recovery never happened in isolation. It was necessary to raise her head, look people in the eyes, and so begin to defeat those regrets and secret shame which make alcoholism so very much a disease of isolation.

Thus Janet's lifelong self-reliance was to be of no use to her in this process. It was difficult, and involved a couple of false starts, but eventually she found another woman she could relate to who would serve as her "sponsor" and help support and guide her through the steps of recovery.

* * *

The sober days began to mount up, but Janet still was plagued with intense resentment whenever she thought of the accident. When she was driving one day, glanced in the rear-view mirror and saw Shane in the car behind her, Janet's heart beat fast.

Janet grimaced as she recalled her feelings. "I looked over at the glove box and was glad I didn't have a gun in there, because I swear, I would've wanted to pull it out and take a shot at her!"

In the smallish community of Ponte Vedra Beach, it was difficult to keep from running into her nemesis. While shopping in a hardware store one day with Kathryn she spied Shane working the checkout register. Kathryn even approached the manager to let him know what an unsavory character he had working for him.

Nightmares about the accident, and about taking revenge, plagued Janet's sleep. In recovery, resentments that are held on to can burn like fire. Giving them up can take time and effort.

It took about three months for Janet to feel comfortable enough in a meeting to share. She was still depressed, emptied out from everything that had happened to her. Most of all, she was to a degree still filled with feelings of worthlessness and failure.

She spoke to her sponsor about her difficulty in talking. "It's funny, I thought I was so used to being "center stage" and in front of people, but this is really hard!"

Janet's sponsor considered. "Yes, but all of that was a kind of acting. When you share from the heart, it's a whole different story."

Nor was her journey to sobriety to be a straight path. More detours lay ahead.

* * *

In early recovery, the alcoholic is often told he must change "old playmates and old playgrounds," for they can stimulate the old cravings. For Janet, the stumbling block was her continuing open-door policy for erstwhile drinking buddy Maxine. Her friend still wandered over, wineglass in hand, needing to talk.

It went against Janet's loyal nature to shut the door on an old friend. Sometimes the two went over to the local American Legion, where Janet had enjoyed the comraderie and the jovial, boozy atmosphere in the past. Now, she wouldn't drink, but would concentrate on shooting pool while Maxine schmoozed at the bar.

Finally, the inevitable happened. Maxine went back to her house one night for a refill, and came back with not one but two drinks in hand. Janet (for no real reason, except that her defenses had been worn down) accepted the glass. That night she drank until she blacked out.

In the aftermath of her secret slip, shamed silence prevailed even as Janet continued going to her meetings. Finally she became so uncomfortable she shared at least part of the truth with her sponsor Julie. She told about Maxine's constantly drinking around her.

Julie registered her sponsee's discomfort and finally gently asked: "Janet, did you drink with her?"

"No!" Janet couldn't bear to have her dear friend know how she had failed. A secret shame burned deep in her heart.

In a meeting a few days later, the subject happened to be "relapse." Janet felt like God was tapping her on the shoulder, but also that she was on the hot seat. Her hand shot up, seemingly of its own volition. She met the steady gaze of her sponsor and said, "I lied to you, Julie."

The truth hurt, but it was freeing too as it poured out. Instead of judgement, Janet saw understanding and compassion in the eyes of her fellow recovering alcoholics. She was beginning to find that the rewards of sobriety were well worth the cost in false pride. As she let more people know more about the real Janet, people responded to her with greater warmth and intimacy.

Back on track, Janet continued to expand her recovery friendships, enjoying lunches and activities with her new friends. She was returning, as well, to the pastimes she loved. She began to golf once more, although the shoulder injuries severely limited a backswing that used to be the envy of others on the course.

It really hurt in the beginning. She had to take pain medication before she played. It hurt, too, to have to lower her own standards for her performance. Janet was always used to playing competetively, even with herself. At first, she would come home and just cry.

There was an important lesson here, she found. She had to change her whole orientation from that of competition to playing simply for love of the game. Janet reached a new level of appreciation for the process itself, for the gift of being out on a beautiful green swatch of God's earth on a warm Florida day beneath a cobalt-blue sky, playing a game she loved. She was finding that there was indeed life in sobriety. And that life was far richer than anything she had known before.

* * *

Maxine, however, was still on the scene. Janet just couldn't seem to keep the door closed to her friend. One day Maxine came up with a diabolically ingenious plan. Why didn't she and Janet set aside just one night a month, and allow themselves to drink as much as they liked for that one night only?

The plan appealed to the remaining part of Janet that thought like a true alcoholic. The illusion of gaining some kind of control over the disease fooled her again. She didn't want to ditch recovery completely, but the lure of being able to have her cake and eat it too was persuasive.

The lure proved to be a lie. She tried it. For a while it actually seemed to work. True, she was feeling awful about herself deep inside again, "faking it" at meetings. Then inevitably came the night when Maxine suggested that since she was going out of town and wouldn't be around for "their" drinking night they should go ahead and indulge themselves right now. Again without even pausing to think about it, Janet complied.

They drank from 6:00 in the evening to 4:00 the following morning. Janet passed out, half dressed, on the floor of her bedroom. When she came to she was so sick she couldn't do anything but crawl into her bed, lie there, and moan. *How had this happened? How could she have been so stupid, so weak?*

When she could finally function, she stumbled into her garage and closed the door. She got into her car and thrust the key into the ignition. Then she paused, hand frozen in mid-motion. One little twist to the right, and the engine would start. She could just lie back, close her aching eyes and surrender to the deadly fumes that would begin to fill the warm, dark garage. She could, at last, give up all the struggle.

But stronger than the shame, the self-hatred, and the guilt was the still small voice of sanity, and the beating of a heart that, though bruised, loved life mightily and wanted to live. She thought of all the people she would hurt. With a trembling hand she switched the key to the left instead.

It was over.

* * *

Janet didn't go to a meeting that night, for she was still too shaky. But the next day she was there. Those who were present that day will remember the honesty of Janet's words and the sincerity of the tears that coursed down her cheeks. One big, burly man was moved to tears himself and spoke of how he would always remember her story and how it would help keep him sober. Again, there was no condemnation, only understanding and a sense among the others that "There but for the grace of God go I."

Another program member became, for a time, Janet's roommate and close friend. One day they were relaxing on the terrace when Maxine showed up to chat, carrying a drink in her hand as usual.

Janet's roommate eyed Maxine levelly. "Maxine," she said, "Don't ever come in here with a drink in your hand again."

She didn't. A badly needed boundary had finally been set. Soon thereafter Maxine moved away. Janet continued to pray for her friend, and was delighted to learn that Maxine finally found sobriety in a religious treatment setting. She was even asked to become the center manager. Janet and Maxine got together regularly, and this time the friends shared not booze but the joys of sobriety.

* * *

It was time to take the next step. Janet and her sponsor walked one afternoon along a wooded trail through the marsh in beautiful Guana State Park, bordering the Intercoastal Waterway, and Janet opened up completely.

Looking back, she smiled. "By the end of those two hours, I had let go of everything. I will go to my grave with no secrets."

And in that release, Janet found release also from the hard core of resentments she had been struggling with. It became increasingly possible to make peace with the past, and accept life on life's terms. She could forgive others because she could forgive herself.

29

WINGS TO FLY

Janet found that life in sobriety was full of unsuspected blessings. She established friendships with other recovering women of all ages. At last, Janet had found an "in group" with whom she felt increasingly at home. Soon she was even more involved as she became Treasurer of a regular daytime meeting and Sponsor, herself, to a newcomer. It was still hard for her to talk in meetings, but what she shared was always heartfelt and often funny.

Jane's vibrant personality acted as a natural magnet for younger women attracted to her vitality, love of life and passions such as turtling or golf. An excellent listener, she sometimes felt guilty that so many of her friends turned to her with their problems

"I don't really have any problems myself today," she marveled.

* * *

The wide world still beckoned. Janet had longed for years to see Africa, home of her beloved lion. In 1988 she and Kathryn took the trip of a lifetime, a 21-day African safari.

In Kenya, the two sisters were thrilled to be lodging beside a watering hole where a great variety of animals came to drink in the evenings. The hotel attendants or guests would ring a bell, even in the middle of the night, when there was something to see. The sisters got to spy on hippos, elephants, even a tiger.

In the wild, they watched cheetahs and lions stalk their prey. Males would lie resting on the sidelines while the females raced after the fleet gazelles, bringing them down with a mighty leap and breaking their necks. Leisurely, the male would stroll over and begin to eat. When he had eaten his fill some 15 or 20 minutes later, the female would feed. Then came the hyenas, and finally the vultures, until, in the great economy of nature, only bones were left. The sisters saw five such kills from their safari van.

At one primitive site they had to go downstairs to the "outhouse" when nature called. When they came back to their room they found the door slightly ajar.

Tentatively, Janet pushed the door wider. They had visitors. Three orangutans had climbed up onto the porch, snuck into their room and were nonchalantly inspecting the premises. Janet ran at them waving her arms, and slammed the door emphatically behind the unwelcome guests.

At dinner another night the guests were getting to know each other around the table.

"And what do *you* do?" someone asked. When Janet shared about Kiddie Kollege, she was startled when a 22 year old girl leaned across the table toward her, eyes twinkling.

"Miss Janet, I'm a graduate of your school!"

Janet found every day to be a new adventure, from a hot air balloon ride (they landed in the same field where *Born Free* was filmed) to a cocktail party in the house featured in *Out of Africa.* Janet's favorite hotel offered a stunning view of Kilimanjaro.

* * *

In 2002, the two sisters were invited back to Aberdeen by the Maritime Museum to celebrate the 70th anniversary of the rescue of the Flying Family by *The Lord Talbot*. When Janet and Kathryn entered the Museum for the first time, they were stunned. An entire wall was covered with the 1932 Life Magazine photo of the year, of Blanche, Kathryn and Janet standing on the Greenland cliff, watching their plane sink.

Unlike America, where the news of the moment takes precedence, in the British Isles the past is a living presence. Thus Janet and Kathryn were amazed to discover the very same photo of the Flying Family that the Hutchinsons had presented to the pub in the Aberdeen hotel they stayed at over half a century ago was still hanging there, on the same spot on the wall.

Some twenty descendants of the original *Lord Talbot* crew surrounded the pair at the museum, two daughters of Captain Tom Watson among them. Kathryn actually remembered daughter Chris

Clark being there when the ship docked, and Mrs. Clark remembered the excitement of welcoming the little girls from the States.

Emotion ran high on both sides. The gratitude of Janet and Kathryn overflowed as they paused to honor the strength and courage of the men who, in their rescue, actually gave them the chance to live. Over the next few days, the sisters were called upon to participate in four separate ceremonies in which the rescue was recounted, questions were asked and answered, and certificates of appreciation presented to the descendants of those very sailors who rescued them so dramatically seventy years ago.

Attendees marveled at the fact that after *The Lord Talbot* returned to Aberdeen, skipper Tom Watson was fired and the entire crew went without wages because the ship returned with the Flying Family instead of a boatload of fish. Undeterred, Watson had gone on in 1935 to engineer another high-profile rescue off the Greenland coast, this time of explorer Martin Lindsay and three of his crew.

Both Janet and Kathryn have framed paintings by a descendant of a crew member depicting Angmagsalik today, greatly changed from the tiny village of seventy years ago.

* * *

Janet's continuing involvement in WASP affairs is a link that keeps her vitally connected to that happiest of times in her life. In 2002 Nancy Parrish, daughter of a WASP, created the *Wings Across America* project in an endeavor designed to document WASP history and bring awareness of this part of American history into the public spotlight. Nancy interviewed some 110 surviving WASP veterans, including Janet, and produced a *Wings Across America documentary*.

The WASP had been denied recognition for some 35 years when, in 1977, Public Law 65-202 granted them veteran status at last. A groundswell of pride and interest began. The records that had been sealed as "classified" were opened, and American history could at last make the WASP part of its story. School texts and written histories of WWII had largely ignored the very existence of the WASP up to this point, since access to the information did not exist.

Brought out of the shadows at last, WASP veterans began to plan reunions and get-togethers. As the population of surviving WASP aged (today, none is under the age of 80) awareness grew that this sisterhood must be honored.

* * *

Texas Women's University in Denton, Texas houses America's official WASP exhibit as part of its Women's Studies program. An extensive collection of WASP history, memorabilia and artifacts are on display, including a bigger-than-life statue created by WASP Jewel Pfeifer Estes. A bronze plaque honors the 38 WASP who gave their lives in service of their country.

In 1994 an impressive WASP memorial and bronze statue were dedicated as part of the National Museum of the United States Air Force at Wright-Patterson Air Force Base in Ohio.

Most importantly to the WASP who served there, a National WASP WWII Museum is being created at Avenger Field back in Sweetwater. Their stated mission is:

To educate and inspire every generation with the history of the Women Airforce Service Pilots, the WASP of WWII, first women in history to fly America's military aircraft and who forever changed the role of women in aviation.

TheWASP/Wings Across America website at www.wasp-wwii.org, supplies a plethora of information, history, photos, and links to WASP doings past, present and future.

Janet's own contributions to keeping WASP history alive are farreaching. In 2003 she was flown to New York City for a special television broadcast celebrating the 100th anniversary of flight. She was interviewed about her WASP experiences and taken around the city for a tour, revisiting scenes from the history of the Flying Hutchinson days and providing commentary. In the North Florida area, she has appeared on television and in numerous newspaper and magazine articles.

* * *

Janet's favorite reunion was held in Sweetwater. It was like walking back through time as she reviewed the familiar sights at Avenger Field. Old faces, remembered voices seemed to live again, just as the scent of rich Texas dust filled the air. Reunions, however, were ever bittersweet. Reunions are a reminder that when the last WASP is gone, only history will retain the memory of this unique page of our collective history. Each year, Janet has seen WASP ranks diminish. None of her roommates are still living, but a few good friends survive.

It is different with the other ongoing branches of the armed forces. New men and women arise to take up the banner of service.

"When we're all gone, there won't be any more like us coming along. That'll be that," Janet sighed.

Janet dogsledding in
Alaska   2001

Janet & son Robbie
Lake Tahoe, NV 2000

Turtle Patrol Memories

Ponte Vedra Beach, FL

30

EVER AFTER

Janet is a person who needs passionate involvement with life, and turtling provides that. In patrolling the beautiful beaches of North Florida to help preserve and protect the endangered sea turtle, Janet celebrates her love of nature, animals, and friendship. Janet's friends have been drawn into the "net" of turtling by her influence.

She met the head of the local Turtle Patrol and his wife soon after her move to Ponte Vedra Beach. She was instantly intrigued. Turtle Patrol volunteers walk an assigned stretch of beach on a regular schedule, usually in pairs. Their purpose is to identify sea turtle nests, post signs for protection, keep watch for hatching babies, and keep count and statistics which are then reported to the state.

Janet looks forward each summer to the return of the nesting season, from May through the middle of September. Her section of Turtle Patrol covers four miles of beautiful Ponte Vedra oceanfront, a span of white sand beach protected by rolling dunes and waving sea oats. The once-wild stretch is now dotted with multi-million-dollar estate homes.

* * *

The indigo sky is still dark but beginning to pinken between the dark purple clouds skimming the horizon, as the turtler's day begins. Only Venus is still visible, hanging low and still bright in the Eastern sky over the Atlantic. A cool breeze rustles through the sea oats rising from the humpbacked dunes, belying the promise of summer temperatures to come that will probably hit 92 degrees before lunchtime. A single bird begins to call. Cicadas sing from the surrounding trees. Mosquitoes hum. Volunteers aim to hit the beach before sunrise, just as the light is sufficient for vision. A sharp lookout must be kept for the telltale tracks that will lead searchers to a new nesting site. They learn to identify the wide, tractor-like markings the big reptiles leave in the sand as they wend their way from ocean to sand and back.

An intrepid few early walkers arrive at the empty beachside parking lot, many with eager dogs straining on leashes. A couple of men, on foot, meander down to the beach. We leave one car at one end of the four-mile walk, and drive together in another to the starting point. As we pass through the gates of the exclusive country club community where the patrol stretch begins, Janet is waved in by the friendly guard who knows her well.

Armed with yellow posting signs, a clipboard, and a hammer, we traipse down to the flat part of the beach where walking is easiest on packed sand. Some days the beach, churned up by a storm or high tides, is rough and uneven. Then the walk feels twice as long. This morning, the beach is hard and broad - good walking.

As the sun clears the horizon, a wide band of molten gold travels from the heart of the bronze orb over the pale water to break with the swelling wavelets on the beach. Dolphins, black against the lightening ocean, rise and fall in pursuit of a school of fish just beyond the breakers. Seldom have we seen them so close to the shore. An osprey soars overhead, clutching a wriggling silver fish in his talons. The beach is beginning to stir with human presence as well. Walkers, runners, dogs. Janet speaks to everyone and knows many of the "regulars."

As we walk in the warming sun, Janet keeps up a running commentary to my questions. I listen to the fascinating saga of the sea turtle.

* * *

Turtle Patrollers stalk the beach area just east of the dunes but beyond the tide line that is favored for nesting by the three most common types of sea turtles that lay their eggs in this section of North Florida. New nests are obvious, signified by the impossible-to-miss tracks leading out of and back into the sea. The tracks are unmistakable, distinct and tread-like; they form a huge V coming in and out of the ocean, the turtle's flippers creating an undulating pattern in the sand that almost looks tractor-like.

When the mother turtle finds her nesting spot, usually between 10 PM and 2 AM at night, she first wriggles her body back and forth to create a shallow body pit. Then she begins to dig with her powerful flippers, troweling the sand out at a furious pace. As she does this, she weeps, or at least her eyes flow with water. She is releasing the salt water from her body through her tear ducts. She deposits her eggs, usually from 50 to 180, covers them back over with sand, and then, exhausted, makes her way back down the beach and into the waiting waves. Her part in this life drama is over for now. Females may lay eggs 3 to 4 times in a single season, then skip several years. .

In North Florida, Loggerhead turtle nests are most numerous, some 85 thousand having been identified in a years' time. Green turtle nests come next, followed by the rare Leatherbacks, the largest, weighing in at 13 to 15 hundred pounds. A Leatherback track is unforgettable, looking somewhat as though a Sherman tank has ploughed its way across the wet beachfront.

Sometimes lucky patrollers glimpse a turtle in the early dawn light making her weary way back down to the sea. If a mother turtle does see a human, she ignores them, completely focused on her task.

(In Jupiter Beach, Janet saw a turtle laying her eggs just one foot away from an observing group. She was totally oblivious. Out of water, a turtle does not see or hear well.)

Once a volunteer has found a new nest, determining where to put the official sea turtle nest sign (designed to identify it and hopefully protect it from molestation), requires care. It shouldn't be placed over the center where the eggs are but in front of the outer edges of the nest.

Occasionally a turtle has made her nest too close to the tide line and the nest itself must be dug up and the eggs moved, or else it will be washed out by the tide before 9 AM. After approximately 12 hours in the sand, the embryo within the egg becomes increasingly susceptible to damage or death during relocation (another reason why turtlers need to get out there early).

Moving the nest itself has to be done at night. Once the nest is dug up, the really tricky part begins. The eggs must be carefully picked up, put in the volunteer's bucket, and placed in the "new" nest further from the tide line in exactly the same position in which they are lying. Extreme care must be taken in order for the baby turtle to continue to develop properly within the shell.

Now the watch begins. It is obvious when nestlings have hatched, as the mound is found dug up and disturbed, and long trails lead down to the sea.

The eggs, which resemble ping-pong balls and feel like a hardboiled egg, incubate for an average time of 55 to 60 days inside the nest. As they hatch, the emerging babies struggle up through the sand to the surface, beginning their fierce fight for survival. Although the lifespan of a sea turtle can reach 100 years, only one in 1,000 will live to the ripe old age of 20, or mature egg-laying age. The struggle itself up through layers of collapsing sand, and the predator-fraught journey down to the edge of the sea makes the baby turtles stronger. If they gain the water, they are still faced with a three-mile swim out to the point where the specific seaweed they need to sustain them grows.

Three days after the turtles have hatched, the volunteers dig up what is left of the nest. They need to make a count and report the status of the nest to the state. They supply statistics in each of 6 areas: the number of empty shells, the number of live hatched turtles, dead hatched turtles, the "pips" that get half of the way out of the shell yet are still alive, the dead pips, and the intact shells (eggs that failed to hatch).

Any remaining live pips, who hatched but didn't quite make it out on their own, can be helped. The volunteer takes the little turtle home if it is too weak to struggle down to the sea on its own. Janet keeps them in the dark, in her laundry room, until the next night when it is released. It's important to put the little creature at the TOP of the beach to start his long walk (again, for strengthening.) When he gets to the shoreline, he is usually pushed back by the waves eight or nine times, and then disappears.

On a form volunteers designate exactly where each nest is located. In Janet's territory, there are huge and elegant homes situated along most of the beach. Their private decks and walkways can extend right down into the dunes. Several sea turtle nests have been found nestled right against a wooden stairway. Family dogs running free, curious children, careless beachgoers can all destroy or harm sea turtle nests. Since in this section of North Florida almost all beachfront is developed this infringement on natural habitat explains why North Florida has relatively few sea turtles.

The turtles fight a valiant battle to survive not only natural predators but human ones. Shrimpers scoop up the sea turtles in their nets, inadvertently. Hungry raccoons can dig up a nest with their handlike paws and feast on the eggs. Worst of all are the humans who deliberately destroy or vandalize eggs. One elderly man was arrested seven times for stealing turtle eggs from nests and selling them. The signs plainly state that there is a $1,000 dollar penalty for disturbing a sea turtle nest, as well as possible jail time.

The "law" of Turtle Patrol states that volunteers should let nature take the lead, and not interfere. If a baby is having trouble making it, they are supposed to be left alone to survive or die as nature dicates.

At the end of the season, the local Turtle Patrollers have a big party to celebrate the end of another year.

As Janet continues to walk the pristine beaches in the dawning days before her, she keeps a watchful eye on the abundant evidence that life renews itself perpetually. Perhaps the cycle she is privileged to witness reminds her of her own life journey as well. Certainly, it is the time at which she feels most at peace with creation and Creator.

* * *

In a country that worships youth and shies away from the idea of aging, Janet's vital life today is evidence that attitude is more important than fact. She continues to leave younger friends panting in her wake, whether on turtle patrol, the golf course, aerobics class, or simply in her continuing enthusiasm for what each day may bring. Janet has never needed an alarm clock to wake up in the morning.

She continues to eat mainly raw foods. Uncooked meat and fish, slivers of raw organ meats from time to time, salads and unprocessed foods make up most of her diet. From the time of sharing snippets of Governor's food she preferred to eat this way, which may account in part of her extraordinary youthfulness and health. She has never had a cavity.

"I've really never been so happy," Janet states, smiling, looking out over her lushly tropical back yard. A pond stands behind the property, home of Gus the Gator and Beauregard the white egret. Red-winged blackbirds, pushy jays, glossy young crows and a plethora of songbirds squawk and dart around the bird feeders that stand beyond the large screened patio windows. Janet seldom uses the air-conditioning, preferring instead to leave her sliding doors open to the Florida air, in all kinds of weather. Where she lives, there is still enough "wildness" in places for a friend to see a Florida panther early one morning, leaping across the highway in front of her car.

Those lucky enough to come within Janet's orbit today feel instantly welcomed and intrigued. Her answering machine, instead of the usual boring outgoing message, greets callers with a corny joke that is changed weekly. A large plaster turtle crawls up the stucco wall facing her front walkway, and ceramic turtle wind chimes greet visitors. A big pink flamingo, lit up with sparkling lights, nods his head to welcome newcomers in the foyer.

In Janet's home, the walls are lined with pictures from her past, her passions, her loves, both of yesterday and today. The Flying Hutchinsons, Janet and Gary Cooper, Governor the Flying Lion sitting at a typewriter, Janet as a beaming young woman of 19 in her WASP uniform. Janet and Kathryn in Africa and Alaska, Janet and Robbie on vacation, Janet and turtling friends with a huge Leatherback turtle on the beach.

Stacks of photo albums and a lifetime's collected memorabilia testify to a life of incredible richness and variety. Every available nook and cranny is filled with Janet's collection of dolphins, elephants, horses, and, naturally, turtles.

Yet it is *today* that most involves and excites her. She is vitally interested not only in what is past, but what is to come. Happiness comes easily. When asked when she is most content, Janet's answer is surprising.

"Just when I'm riding in my car, looking around and seeing all the beauty, appreciating it." As long as Janet can enjoy intimate communion with the natural world, she is content.

She continues to cherish her solitude. Living next door to sister Kathryn and relatives, in close touch with son Robbie, she cherishes friends and family yet loves living alone. "I'm never lonely, and I'm never bored."

* * *

There is a cemetery, set among spreading live oaks and bordered by wild Florida scrub, not far from Janet's home. The setting is lush with life. Pheasants and their babies trail through the underbrush. An osprey flies over clutching a silver fish, heading for a mossy nest high in a lightning-scoured pine. Mockingbirds dart and call. Water splashes from a nearby fountain. It is a place of peace and natural beauty, where the overdevelopment fast engulfing the Palm Valley wilderness will not be allowed to encroach.

Here is a green plot with headstone already in place. Janet designed it herself. Carved on its granite surface is an airplane, a horse, and two books (representing her father's books on the Flying Family's adventures). Realizing she had left out turtles, Janet was given one to place atop the grassy mound. It lights up at night.

She is still looking for a stone lion.

Janet usually visits the cemetery once a week. She finds a sense of peace and solace there, knowing herself to be a part of the natural order of things. Here past, present and future melt peacefully into the beauty of Now. Many can't understand this ritual, but for Janet it is just a normal part of her life.

The blue eyes shine as she describes it. "And you know," she adds, "Every time I drive past it, I wave!"

* * *

Sobriety, sharing, and an ever deeper appreciation for life in all its aspects have enriched Janet's life immeasurably in these last few years, and in the process of her own transformation the circle of her influence has widened and deepened. Her life story is a remarkable one, but knowing the woman *behind* the story is even more remarkable. When we encounter a genuine lover of life, something deep within us responds. We grow. And Janet is such a one.

THE END